CW00326305

The Grub Street Dictionary of International Aircraft Nicknames, Variants and Colloquial Terms

John Horton

GRUB STREET · LONDON

Published by
Grub Street
The Basement
10 Chivalry Road
London SW11 1HT

Copyright © 1994 Grub Street, London
Text copyright © John Horton

British Library Cataloguing-in-Publication Data
A catalogue record is available on request from the British Library

ISBN 0-948817-75-7

All rights reserved. No part of this publication may be reproduced, stored in a
retrieval system, or transmitted in any form or by any means, electronic,
mechanical, photocopying, recording, or otherwise, without the prior permission
of the copyright owner.

Edited by Daniel Balado-Lopez
Typeset by Pearl Graphics, Hemel Hempstead

Printed and bound in Great Britain by
Biddles Ltd, Guildford and King's Lynn

Contents

Acknowledgements iv

Introduction v

The A-Z Dictionary 1

APPENDIX 1
Allied Reporting Names for WW2 Japanese Aircraft 153

APPENDIX 2
NATO/ASCC Reporting Names 160

APPENDIX 3
Formal British Military Naming Systems 168

APPENDIX 4
Series Names 173

APPENDIX 5
Phonetic Alphabets 175

APPENDIX 6
Nicknames of US Navy Aircraft Carriers 177

Glossary 178

Index 180

Acknowledgements

In compiling this book, I have received invaluable help from many individuals and organisations, some of whom I perhaps failed to thank adequately at the time. A big Thank You then to Agusta S.p.A, Air Britain, Jean Alexander, Dave Allen, American Airlines, Dr Surendra Arora, Al Becker, The Beaufort Aircrews Association, The Blenheim Society, The Chipmunk Club, Peter Davis, Dept of the Air Force (Washington), Mike Donson, Dr Michael Fopp, The Indian High Commission, David Legg, John Maynard, Peter J. Marson, D. Menard, R.J.W. Merison, Tony Merton-Jones, Sean O'Brien, The Shackleton Association, Terry Sykes, the Taipei Representative Office in the UK, Peter Thompson, United Airlines, The USAF Museum, The Vulcan Restoration Society, David Watkins, John A. Whittle.

Special thanks to Richard Clarkson, to whom I still owe several pints of beer, and to Les Leetham, who has a much better story to tell. Also to the Training Captain of the late, lamented Dan Air (whose name I have sadly mislaid) who took the trouble to answer my oddball query on the very day his airline ceased to exist.

But more than anyone else I must thank my Mother. Her practical help was enormous, but even more important was her faith and encouragement, without which I would simply have given up in despair.

Introduction

In recent times the naming of aircraft has perhaps become something of a lost art. Where there were once such evocative and memorable names as Spitfire, Black Widow, Comet and Stratocruiser, all too often now we have sterile titles like Eurofighter, Regioliner and Heliliner, and ever more frequent raids into the history books to produce names like Thunderbolt II and even Globemaster III. Many types are not even named at all, just given anonymous designations like B767-300ER and BAe RJ70.

This is unfortunate, because there are still many people who care about aircraft names, their histories and traditions, just as there always have been. The recent controversies over names for the USAF's ATF candidates (the Lockheed YF-22 and the Northrop YF-23), for example, hark back to the Gloster Meteor affair of WW2, and even the Sopwith Camel and Pup episodes of WW1, and all illustrate that a flourishing sub-culture of names and nomenclature can still operate, despite the lack of interest within officialdom and big business.

This alternative culture is perhaps at is most inventive and lively in the form of nicknames and other colloquial terms attached to aircraft. In some sections of the US forces, for example, it is virtually *de rigueur* to use an acceptable nickname, unless you wish to stand out as the FNG (... New Guy) of the group; which is why you may hear servicemen talk amongst themselves of Vipers or Warthogs, but will rarely hear mention of Fighting Falcons or Thunderbolt IIs.

Nicknames can originate in a multitude of ways, ranging from the aircraft's appearance, flying characteristics, construction or design features, safety record or references to contemporary people or events, rhyming slang or plays on the formal name or designation; or of course from a simple diminutive form of the official name. Such names usually originate with the people most closely involved with the aircraft concerned, and as such often reflect aspects of the aircraft that are otherwise overlooked or glossed over.

'Nicknames' in this context refers of course to colloquial or informal terms, and is not used in the occasional American sense of 'Official Nicknames' used in connection with such wholly formal titles as, for example, Phantom or Eagle. Furthermore, nicknames here are those used for a type or marque of aircraft as a whole, rather than individual machines within that marque; thus, whilst you will not find references in the following pages to the B-17 Flying Fortress 'Sally B' or the HP Halifax

v

'Friday the 13th', you will find 'Fort' and 'Hallybag'. Nicknames are identified in the book in italics. Cross-references in the text to other entries are given in bold, but if it is a cross-referenced nickname it will be in bold italics and because the adjectives 'Old' and 'Flying' tend to be added or dropped indiscriminately, these are discounted for purposes of alphabetical listing; thus, for example, 'Old Shaky' will be found under 'S'. Some types of aircraft appear under more than one heading, but full cross-references are provided in the index.

Many unusual names also originate with what may be termed secondary operators or manufacturers. Many airlines have used their own class names or marketing titles, ranging from Imperial Airways' Scipio class Short Kents to today's Singapore Airlines Big Top and Mega Top Jumbos. The Military, too, often adopt their own names for foreign aircraft, especially if they are built or upgraded locally. Names of conversions and relatively minor variants are also included, and whilst a degree of selectivity is unavoidable in the latter case, such well-known variants as the Avro Manchester/Lancaster/Lincoln are omitted (or touched upon only briefly) as these are already well documented elsewhere. Perhaps less well-known though are Hawker Siddeley's AEW Toucan project, the Shorts/Lockheed Georgia Bel, or the aptly named Colossal Guppy proposal based on the Boeing B-52 bomber.

Also included are names that were once considered for an aircraft type, or even formally adopted and then dropped. For example, if things had worked out differently, the RAF in the sixties and seventies could easily have been flying Albion, Harpoon and Trenchard bombers, Hurricane, Excalibur and Spectre fighters, and Tercel trainers.

This, then, is a brief outline of the kind of names to be found in the following pages, although of course many refuse to be pigeonholed quite so conveniently. Therefore the book should perhaps best be regarded simply as an anthology of lesser known aircraft names, and one that will hopefully provide an unusual insight into the international culture and language of aviation.

John Horton
'Gwyndy', May 1994

Aardvark

Few aircraft have attracted such a wide variety of nicknames as the General Dynamics F-111. During its troubled development programme in the mid-sixties the American Press dubbed it *McNamara's Folly* (after the Secretary of Defence) *Lemon, Seapig* (the overweight F-111B naval interceptor) and *Flying Edsel* after the disastrous Ford car, but over the years both the reputation and the names have changed dramatically. In south-east Asia it was known as *Whispering Death* to the Pathet Lao (because at low level it usually arrived without warning) and as *Snoopy* to US forces, followed in later years by *Lizard, Switchblade, Bullet Bomber* (SAC's FB-111A) and even *Cadillac*. But of course the most enduring nickname of all has been *Aardvark*, first coined by instructor pilot Al Mateczun in 1969, and occasionally translated into its English language equivalent of *Earthpig*; the Royal Australian Air Force variation *Pig*, however, refers more to its *modus operandi* than its appearance, with that long nose grubbing in the dirt.

Though the One-Eleven had more than its fair share of nicknames, the only variant to be given a formal name was the Grumman-developed EF-111A Raven electronic warfare aircraft, also known colloquially as the *Electric Fox* and *Spark Vark*. In view of such American titles as Fighting Falcon and the fiasco over naming the Rockwell B-1B it may be considered fortunate that the standard F-111 models were never given a proper type name, although it also begs the question of what the RAF might have called the cancelled F-111K.

Able Mabel

Phonetic style nickname (pure phonetics would have been Able Mike) for the Martin AM-1 Mauler, a short-lived US Navy type of the immediate post-WW2 period that had the misfortune of having to compete with the legendary *Spad*. Others claimed the AM designation really stood for Awful Monster, but this never really amounted to a proper nickname.

Academe

Rarely used name for the US Navy's Grumman TC-4C, a training adaptation of the twin-Dart Gulfstream 1 transport fitted with the radar and associated systems of the A-6 Intruder attack bomber; nine were ordered in 1966. Commuter was the obvious title attached to a 1979 proposal by new owners Gulfstream American to stretch the original turboprop business aircraft by 10ft to produce a 38-seat commercial transport; just six were converted as GAC 159-Cs and plans for a new-build programme fell through.

Ace Maker

Contrived but nonetheless deserved title bestowed upon the sturdy and reliable Grumman F6F Hellcat fighter which, although lacking the glamour of the contemporary F4U Corsair, accounted for nearly three-quarters of the US Navy's air-to-air kills in WW2; over 5,000 in less than two years' operational service. The Royal Navy version was known briefly as the Gannet before the standardisation of American and British names in 1944.

Ace of Spades

The OED defines a javelin as a 'light throwing spear', so it was perhaps a questionable title for the RAF's massive all-weather fighter of the late-fifties/early sixties. In the early days of

1

rivalry with the DH 110 (forerunner of the Sea Vixen) de Havilland personnel apparently called it the *Trug* (a kind of wooden basket) but the exact reason is unknown. The unusual tailed-delta planview of Gloster's last aircraft also earned it such names in RAF service as the *Flying Triangle*, *Flat Iron* and *Ace of Spades*, while *Harmonious Dragmaster* came from the distinctive whine of the twin Armstrong-Siddeley Sapphire engines and the 'organ-like' note of the airflow over the gun ports.

Achilles
A rather unfortunate original choice of name for the Armstrong Whitworth AW55 airliner, in view of the fact that it was problems with the Mamba turboprops (and the promise of the Viscount) which ultimately put paid to a basically sound design. The name Avon was adopted in March 1947, only to run into objections from the Ministry of Supply who wanted river names reserved for Rolls-Royce turbine engines, but by the time of its maiden flight in 1949 the AW55 had been given its final name of Apollo.

Adam
Phonetically inspired nickname in its native Sweden for the initial model of Saab's famous Draken fighter, the J-35A, followed in due course by the *Bertil, Caesar, David, Erik, Filip* and *Johan*. The Draken name itself is usually translated as simply 'Dragon' but is strictly '*The* Dragon' ('Dragon' alone is 'Drake' in Swedish, without the final 'n') and was first applied to the 1951 Saab 210 built to test-fly a roughly half-scale version of the famous double-delta wing; with the advent of the ultimate J-35 fighter series the 210 became known as the *Little Dragon*. In Finland the Draken

is sometimes referred to as the *Sigurd* after the hero of Norse mythology.

Adolfine
It was reputedly none other than Adolf Hitler who first thought of relaying the Olympic torch from Greece, and after the 1936 games in Berlin the Führer consented to his name being used unofficially for the Messerschmitt Me 261, the sole purpose of which would be to fly the torch on a record-breaking flight to the 1940 Tokyo games. Despite early enthusiasm the project necessarily took a back seat to Bf 109 production and the first of three aircraft did not fly until 23 December 1940, by which time, of course, the project, the games, and the world itself had all been overtaken by events.

Aguila
A Chilean Air Force upgrade of the classic Hawker Hunter fighter, carried out locally by ENAER, the Aguila (Eagle) incorporates a new US-style cockpit layout and fin-mounted Caiquen II Radar Warning Receiver (RWR), amongst other improvements.

Airbuskie
Occasional Western reference (cf *Concordski*) to the Soviet Union's first widebody airliner, the Ilyushin Il-86 (NATO-Camber) although when it first flew in 1976 Airbus Industrie was still fifteen years away from its first four-engined A340.

Air Force One
Strictly speaking, this is simply the callsign of any USAF aircraft carrying the President, but it was perhaps inevitable that the name would rub off informally onto the Boeing C-137s (707 variants, formerly designated

VC-137s) and VC-25As (747s)
operated for that purpose by the 89th
MAW at Andrews AFB, Maryland.
The term has also largely supplanted
the old *Flying Whitehouse* name for
what are undoubtedly the most 'seen'
aircraft in the world. Similarly,
Marine One refers to the USMC Bell
VH-1N Hueys (now retired), Sikorsky
VH-3A/D Sea Kings, and VH-60A
Blackhawk Helicopters used for
flights in the Washington area and
flights within other countries. *Air
Force/Marine Two* is used for the
Vice-President's aircraft, and so,
pedantically, 'Two' became 'One' the
moment Lyndon B. Johnson was
sworn in aboard VC-137C 62-6000 on
23 November 1963, less than two
hours after Kennedy's assassination;
LBJ also coined the term *Air Force
One-Half* for the smaller Lockheed
VC-140B JetStar. Other variations on
the theme have included the callsigns
'Peanut One' during Jimmy Carter's
term of office, and 'Freedom One' for
the VC-137C which returned the 52
American hostages from Iran in
January 1981. At election time,
aircraft carrying the Press on the
campaign trail are sometimes known
as *Zoo Planes*, and *Flying Taj Mahal*
has been used by the media to
describe the opulence of the VC-
25As, which could be just a touch
disconcerting, considering the original
is not actually a palace but a
mausoleum.

Air Microbus
Name used briefly by the Soviet
Aviaexport agency in its vain attempts
to market the Beriev Be-30 (NATO –
Cuff) seventeen-seat feederliner in the
West before abandonment in favour
of the An-28 (NATO – Cash) in 1970.
The more powerful Be-32, reported as
a new model at the time of its

appearance at the 1993 Paris Air
Show, was in fact flown originally in
the mid-seventies.

Airwolf
Sadly, the supersonic Airwolf
helicopter belongs firmly in the realms
of Hollywood fantasy, and the hi-tech
machine flown by Ernest Borgnine
and Jan-Michael Vincent in the TV
series of that name is in fact a standard
Bell 222 converted for the part by
Jetcopters Inc. Modifications include
dummy armament and jet engines in
fibreglass, and the high-speed effect is
achieved by shooting the sequence at
a slower film speed, but showing it at
the normal rate.

Ajax
Ajax cannot legitimately be
considered the original name of the
RAF's Atlas Army Co-op biplane, but
for reasons unknown Armstrong
Whitworth used the title for two or
possibly three virtually identical
prototypes in 1925. The one-off Aries
was a 1930 attempt at a more easily
maintainable Atlas, with reduced
rigging and more fuselage access
panels.

Albion
Following its show-stopping debut at
Farnborough in September 1952
(rolls, turns within the airfield
perimeter, and all with less than three
hours on the clock) there was
considerable speculation on a possible
name for the Avro 698 delta-winged
jet bomber. *Flight* magazine ran an
article suggesting Apollo, Assegai,
Avenger, and their favoured option
Albion, but the following month the
V-Bomber class was born, and the
second of the trio went down in RAF
history as the Vulcan. Nicknames are
surprisingly scarce, with *Tin Triangle*

3

and *Flat Iron* being largely restricted to the layman and nearly as bland as the clichéd (and inaccurate) description 'Bat-Like', so beloved by the media. A proposed 600 mph airliner development with a lengthened fuselage seating up to 113, the Avro 722 Atlantic, failed to materialise but can still be seen in model form at BAe Woodford.

Aldon
Britain bagged its first Messerschmitt just hours after Chamberlain's radio broadcast on 3 September 1939, the weapon involved being a six-inch nail hammered into one of the tires of the German Embassy's Bf 108 Taifun (Typhoon) lightplane D-IJHW at Croydon Airport; in true 'Dad's Army' style the ferry flight to the RAF ended in a forced landing and the local constabulary seized the unfortunate crew as spies. Altogether about twenty Bf 108s were captured or impressed into the RAF and renamed **Aldon** after BFW-Messerschmitt's pre-war agent. J.H. Aldington, possibly to mask their German origin and to avoid confusion with the Hawker Typhoon fighter.
    Production of the Bf 108 was transferred to occupied France in 1942, and post-war Nord went on to produce it for the French armed forces as the Renault-powered Pingouin. Both prototypes of the tricycle-geared Me 208 were also built at Les Mureaux and the type was developed post-war into the Noralpha, but the bulk of the 200 built were delivered to the military with the title of Ramier (Woodpigeon).

Amerika-Bomber
Unofficial title of the Messerschmitt Me 264, designed to attack the USA's eastern seaboard, particularly New

York, from European bases. Due to changing requirements only the single prototype flew, and its existence was not revealed until July 1944 when a German radio station announced it had been made ready to take Hitler to Japan, had the purge of dissenting army generals failed.

Amiens
Something of an oddity in aircraft nomenclature, **Amiens** was originally applied to the de Havilland DH 10 twin-engined bomber in 1918, but appears to have been quite simply forgotten until rediscovered in Government war records in the fifties. Even now the name is little used.

Anaconda
Soviet test pilots' epithet for the terrifying Lavochkin La-250 interceptor, a vicious beast over 80ft in length that was abandoned in 1958.

Anade
Popular name for the Russian Anatra D recce biplane of 1916; similarly the DS variant took its name of *Anasal* from its Salmson radial engine.

Anakonda
Not to be confused with Polish Mi-2 developments (cf **Kania**) is the all-new PZL Swidnik W-3 Sokol (Falcon) twelve-seat helicopter, first flown in 1979. Variations on the standard transport model include the Polish Navy's W-3RM Anakonda search-and-rescue helicopter, and a dedicated anti-submarine warfare model is projected for 1997 as the W-3U-1 Alligator. An assault version for the Polish Army was test flown as the W-3U Salamanda, but was discontinued in 1991 after problems with its Soviet weaponry, comprising GSh-23 cannons, 80mm rocket launchers, and

AT-6 (NATO – Spiral) anti-tank missiles. The Salamanda was to have been succeeded by the W-3WB Huzar with weapons from South African company Denel, but this was suspended on financial grounds in 1993, as was development of the projected W-3L Sokol-Long stretched transport carrying up to fifteen passengers.

Angelito

A small number of Henschel Hs 123 dive-bomber biplanes served with the Condor Legion in the Spanish Civil War, receiving the popular name of *Angelito* from the Nationalists.

Anteater

Ameisenbar (*Anteater*) and *Pfeil* (*Arrow*) were unofficial names applied by Luftwaffe and Dornier test pilots respectively to the fabulous Do 335 fighter. With its fore-and-aft engines, the 474 mph Do 335 remains one of the fastest piston-engined aircraft ever to go into production.

Apache

After rejecting North American's favoured name of **Apache** for the P-51 fighter, the USAAF did not take up the British title of Mustang until the advent of the Merlin-powered P-51B model; as a result, **Apache** is sometimes used erroneously for the early Allison-powered A-36A ground-attack model, the first of the breed to be ordered for the USAAF's own use. The A-36A was also referred to by front-line units in Sicily as the *Invader*, but as that name had already been pencilled in for the Douglas A-26 bomber, it too remained unofficial. Real nicknames for the P-51 were relatively scarce, but included *Peter-Dash-Flash*, and the simple abbreviation '*Stang*'.

Perhaps the greatest testimony to the thoroughbred qualities of the Mustang was the success achieved by Trans Florida (later Cavalier Aircraft) with their civil Executive Mustang and Cavalier conversions. Between four and six conversions a year were carried out in the early sixties, with twin seats, updated avionics, luggage lockers in the former gun bays, and various fuel capacity options. Following their Mustang II (1967) and Rolls-Royce Dart-powered Turbo Mustang III (1968) counter-insurgency prototypes, the company went on to the ultimate P-51 derivative with the Enforcer, two of which flew in 1971 with 2,535 eshp Lycoming T55 turboprops. Piper acquired the programme and in 1983 built a second pair of Enforcers as PA-48s for evaluation by a somewhat reluctant USAF, but by then the type had less than ten per cent commonality with the original P-51.

Arguably the most radical of all P-51 conversions though was Fort Wayne Air Service's unique *Learstang* racer, fitted in 1988 with the wings and horizontal tail of the Learjet corporate aircraft.

Applecore

Described by Terence Horsley in *Find, Fix and Strike* as a *Gentleman's Swordfish*, the WW2 Fairey Albacore torpedo bomber suffered the indiginity of actually being outlived by the famous *Stringbag*, the very type it was designed to replace.

Flying Aquarium

Nickname of the German Arado Ar 198 recce prototype of 1938, with its extensively glazed underbelly.

Aquilon

De Havilland Sea Venom variant

built under licence by SNCASE for France's Aeronavale in the mid-fifties. A total of 131 Aquilons (Sea Eagle) were built wholly in France, in both single- and two-seat form, powered by a Fiat-built DH Ghost 48 turbojet of 4,840lb thrust.

Argus
Most Fairchild UC-61s, the wartime variant of the Model 24 cabin monoplane, went to Britain where they were known by the name **Argus**. The USAAF introduced the name Forwarder for the UC-61K model, with a Ranger inline in place of the more familiar Warner radial, but in Britain this became the Argus III.

Arrow
Predictable nickname for the experimental Lavochkin La-160 of 1946, the Soviet Union's first swept-wing jet fighter.

Arrowbile
One of the great dreams of aviation, particularly in the US, has been the notion of 'an airplane in every garage', or better still a roadable aircraft, i.e. a machine that could be used as both a car and an aircraft. Few people have come as close to the dream as Waldo Waterman, who married the wings of his tailless Arrowplane to a new tricycle-wheeled nacelle housing the engine from a 1937 Studebaker auto, capable of driving either the pusher propeller or the rear wheels. The **Arrowbile** proved capable of at least 45 mph on the ground and 100 mph in the air, and was convertible in just three minutes by removing or replacing the wings and propeller. Things looked promising when Studebaker agreed to set up a dealership at an on-the-road (or in-the-air) price of $3,000, but rising

costs and WW2 sadly put an end to the project with only about six completed; the design was revived briefly in 1948 as the Franklin-powered Aerobile, but again with no commercial success.

Aspro
In some respects analogous with the American *Jenny*, the ubiquitous Avro 504 biplane was so famous in British aviation that it was usually known simply as the Avro, although this was occasionally punned as *Aspro* in WW1. The nickname Crab appears to date from the thirties, and Anahuac was the title given to a variant of the 504K built under licence in Mexico in the twenties. The limited production 504R (originally the 504J MkII) of 1926 was named Gosport in recognition of the pioneering work in evolving modern training methods by the 504J-equipped School of Special Flying, founded at Gosport (hence also the Gosport speaking tube) by Major R.R. Smith-Barry in 1917.

Ass Ender
Archaic American expression for canard (i.e. tail-first) aircraft, particularly when the illusion of flying backwards is reinforced by the use of a rear-mounted pusher propeller. The name became so rife in the twenties and thirties that several manufacturers made the best of it and shared the joke by calling their designs Ascender; examples include the Granville Brothers' (of Gee Bee fame) experimental lightweight, and the Curtiss XP-55 fighter prototypes of 1943.

AStar
North American marketing name for an optimised version of the Aero-spatiale/Eurocopter AS 350 Ecuriel light turbine helicopter powered by a

6

Textron Lycoming LTS 101 turboshaft in place of the basic Turbomeca Arriel 1B; similarly the name SuperStar is used for the uprated AS 350B2 (whilst still retaining its Turbomeca Arriel 1D1 engine) of 1990, and TwinStar for the AS 355 (military AS 555) Ecuriel 2 of 1979 with a pair of Allison 250-C20s. Elsewhere the series is marketed under the original French title of Ecuriel, although the English translation of Squirrel is more commonly used in the UK; Helibras of Brazil also uses the name Esquilo or Squirrel for licence-built versions of both the single- and twin-engined lines. **Fennec** was the name introduced by Eurocopter in 1991 for military versions of both the single- and twin-engined models. Rocky Mountain Helicopters' conversion based around the installation of a new Allison 250-C30 turboshaft is known as the AllStar.

Atlant
A conversion of the Soviet-CIS Myasischev Mya-4 Molot (Hammer) strategic bomber (NATO – Bison). the VM-T **Atlant** features a new endplate tail assembly to allow piggy-back carriage of outsize loads, notably sections of the Energia space rocket.

Attack Guppy
US Navy nickname for the plump Rockwell T-2 Buckeye jet trainer, particularly amongst aircrew hoping to go on to fly strike aricraft. The T-2 takes its formal title from its native 'Buckeye State' of Ohio, named after a certain species of tree.

Augsburg Eagle
To the man in the street, the Messerschmitt Bf 109 quickly became *the* German fighter of WW2, despite early British attempts to discredit it.

Inspired perhaps by some of the caustic remarks from rivals Heinkel (whose He 112 unexpectedly lost out to the 109), the British Press gleefully overstated teething problems by dubbing it the *Flying Brick* or *Flutterschmitt*. The more worthy and enduring title of *Augsburg Eagle* refers to its birthplace of course, but pilots usually knew each successive model by a phonetic suffix, thus the Bf 109B was the *Bertha* and was followed by the *Clara, Dora, Emil, Fritz* and *Gustav*; the Bf 109K perhaps appeared in too few numbers to receive a popular name, but *Karl* would have seemed appropriate (see Appendix 5). The Bf 109G-1/Trop was also known as the *Beule* or *Bump* after the fuselage fairings over the breech blocks of its MG131 machine-guns.

In its latter marks the once nimble 109 had become rather overcooked as a fighter, but none were as unpopular as the Avia S-199 variant (and two-seat CS-199) built in post-war Czechoslovakia with the wholly unsuitable Jumo 211F engine; excessive wing-loading and a paddle prop gave it some vicious take-off and landing traits, and it was widely condemned as the *Mezec* (*Mule*). Spain had more success with post-war developments, and from 1953 Hispano converted some 143 of the Hispano-powered HA-1109 derivatives into Merlin-engined HA-1112 Buchon (Pouter Pigeon) ground-attack aircraft. Ironically, the quintessential German fighter had begun and ended its career with Rolls-Royce engines (the prototype having first flown with a Kestrel in May 1935) and even the Emils in the 1969 film *The Battle of Britain* are in fact Buchons.

Auntie Ju
Old-fashioned, wrinkled and sedate, the tri-motor Junkers Ju 52/3m, mainstay Luftwaffe transport throughout WW2, was known throughout the Wehrmacht as the *Tante Ju* (**Auntie Ju**) or occasionally *Iron Annie*, while Allied servicemen dubbed it the *Corrugated Coffin*. The fledgling BEA gave the class name of Jupiter to the war reparation Ju 52/3ms it inherited from Scottish Airways and Railway Air Services in February 1947 and withdrew in August of the same year. Almost 450 aircraft were produced at the former Amiot works after the liberation of France as the AAC 1 Toucan, although the name was rarely used.

Autogiro
Title registered by the Spaniard Juan de la Cierva in the twenties to differentiate autogyros (note spelling variation) and gyroplanes of his design or derivation from those of others. Whereas in a helicopter the powerplant is used to drive the rotor (via the transmission) the engine on an autogyro is used for forward propulsion, like a conventional aircraft, and the rotor autorotates or 'windmills' in the slipstream. Cierva flew the world's first successful autogyro, the C4, at Getafe on 9 January 1923, and over the years his British-based company granted licences to Airwork, Avro, Comper, de Havilland, Weir, Westland (all UK), Buhl, Kellet, Pitcairn (all USA), Liore et Olivier (France) and Focke-Wulf (Germany). Development of the autogyro petered out as a result of Cierva's death in the KLM DC-2 crash at Croydon on 9 December 1936, and the eventual refinement of the helicopter.

Avenger
Name coined by Eugene Gilbert for the WW1 Morane type N monoplane with the famous machine-gun deflector plates, shortly after Roland Garros force-landed behind German lines in a similarly equipped Morane type L parasol (not a type N as was once believed). In the Royal Flying Corps the type N was better known as the *Bullet*, particularly those machines fitted with a propeller spinner, and also as the Monocoque Morane in the mistaken belief that it was of monocoque construction.

Avitruc
Chase Aircraft derived the C-123B twin-engined tactical transport from its XG-20 Avitruc glider, and also built the first five powered versions itself; however, when it became apparent that parent company Kaiser-Frazer could not meet its USAF order for 300, the type was taken over by Fairchild in 1953 and renamed Provider.

Baaz
Indian Air Force name (translating as Eagle) for the MiG-29 (NATO – Fulcrum) fighter. The Fulcrum C model, with a deeper fuselage aft of the cockpit, is known to CIS crews as the *Gorbaty* (*Hunchback*).

Baby Boeing
Early nickname for the Boeing 737 short/medium-range twinjet, the smallest of the Boeing family of jet airliners, and familiar to thousands as the backbone of the UK package holiday fleet. The original 737-130 first entered service in early 1968 with Lufthansa, who promoted it as the City Jet, although internally it also became known as the *Pig*; personnel at Britannia Airways knew the 737-

200 more sympathetically as the Funny Little Ugly Fat Fella or *Fluff*.

The largest military order to date has been from the USAF for nineteen T-43A navigation trainer versions of the 200 series, known colloquially as the '*Gator* (from Navigator) or *Strike Pig*. Surveiller is the name used by Boeing for the trio of coastal-patrol 737-2X9s supplied to the Indonesian Air Force from 1982, with Motorola Side-Looking Multi-Mission Radar ('Slammer') in twin parallel fairings on the upper rear fuselage.

Baby Clipper
Informal title within Pan Am for the twin-engined Sikorsky S-43 amphibian, a smaller brother to the four-engined S-42 and roughly in the Douglas DC-2 class. The first Pan Am aircraft to carry the famous Clipper title was the Sikorsky S-40 flying boat NC80V (American Clipper) in 1931. The airline registered the name as a trademark, using it as both a radio callsign and as the basis of individual aircraft names, until its demise in December 1991.

Bacalla
Popular name within American Airlines for the BAC One-Eleven twinjet. Although the name clearly derives from the acronym of the British Aircraft Corporation, many aficionados wince at the mention of a 'Back One-Eleven' and insist on it being pronounced 'Bee-Ay-See'.

Badger
Former name of the 1956 Beech 95 Travel Air light twin, supposedly changed to avoid confusion with the Soviet Tu-16 bomber. In 1960 the French SFERMA organisation modified a single Travel Air to take Astazou II turboprops as the

SFERMA 60 Turbo Travel Air, but the definitive SFERMA 60A Marquis was actually a conversion of the Beech 55 Baron, the parent company's own major derivative of the Travel Air. Despite the increase of over 60 mph on cruising speed only about twenty Marquis conversions were carried out.

More successful, with over 250 conversions, has been Colemill Enterprises' President 600, fitted with the company's usual 300hp Continental piston engines and three-blade props. Colemill also do the more radical Foxstar, a Baron 55 or 58 with the uprated engines plus Hartzell Sabre Blade props, 'Zip-Tip' winglets and numerous systems improvements.

The US Army's instrument trainer version of the Baron is the T-42 Cochise.

Balalaika (i)
Nickname of the Sukhoi T-3 fighter prototype flown in the Soviet Union in 1956. Just as the Balalaika is considered a relatively basic and unsophisticated instrument, so the tailed-delta T-3 was thought to possess a limited aerobatic repertoire.

Balalaika (ii)
Serving with over 40 air forces and standard Warsaw Pact issue throughout the sixties and seventies, the MiG-21 (NATO – Fishbed) fighter followed the usual Soviet practice in not having a formal type name, although the relative simplicity of early models led to the nickname *Balalaika*, and the MiG-21PF fitted with a scaled-down version of the wing for the Tu-144 (NATO – Charger) supersonic transport was referred to as the Analogue. Blue Bandit was the US codename for the MiG-21 in Vietnam, and in India the locally-built MiG-21bis is named Vikram (translating as

'Heroism' or 'Valour', and also the name of a great king of Ujjain).

In the early eighties, China unveiled the F-7M Airguard, an export model of its J-7 II (upgraded MiG-21PF derivative) with a Western avionics suite including a GEC Head-Up Display (HUD). Amongst the buyers was Pakistan, who also funded a joint Chinese/Grumman (US) study into a version with a General Electric F404 engine, lateral intakes and 'solid' radar nose; this was revealed as the Sabre II in September 1987, but was abandoned on cost grounds in 1989. Instead, Pakistan opted for another F-7M development tailored more closely to its requirements – the F-7P Skybolt – with Martin Baker ejection seats and the ability to carry four air-to-air missiles instead of the usual two. Super-7 is the title of a further Chinese/Grumman study along the lines of the Sabre II, but with a weapons system based on that of the Lockheed F-16, either a GE F404 or Rolls-Royce RB 199 engine, and the cockpit hood from the Northrop F-20 Tigershark.

Balzac
Balzac, taken from the famous 'Call Balzac 001' slogan of a Paris advertising agency, was used as a callsign for several of Dassault's '001' prototypes and also adopted more formally for the 1962 Mirage III VTOL conversion. Powered by eight Rolls-Royce RB 108 lift engines and a single Bristol Orpheus for forward propulsion, the Balzac V (for 'vertical') was converted from one of the early short-span Mirage III airframes and was essentially a technology demonstrator for the larger Mirage IIIV. The Balzac V was not rebuilt after its second fatal crash on 8 September 1965, and the Mirage

IIIV-02 (second prototype) was also lost in November the following year. Development was later discontinued.

Bamboo Bomber
Occasional RAF slang term for the wooden de Havilland Dominie biplane, the trainer/comms variant of the classic DH 89 Dragon Rapide light transport; the Dominie title itself (cf **Jet Dragon**) was not bestowed until January 1941, nearly six years after the DH 89 first entered RAF service. Despite the name, the Dragon Rapide was more of a scaled-down DH 86 (see **Diana**) than an improved DH 84 Dragon, and as such it is perhaps fitting that it was often referred to simply as the 'Rapide'. BEA christened its Dragon Rapides the 'Islander' class in 1950, although apart from the flagship G-AHXX 'Islander' itself, the rest of the fleet (uniquely within BEA) were not given names concomitant with their overall class name, being named instead after famous people.

Flying Banana (i)
Royal Naval Air Service nickname for the Franco-British Aviation (FBA) Type B two-seater flying boat, after its yellow-varnished hull and generally 'bent' fuselage shape; most British FBAs were built under licence by Gosport Aviation or the Norman Thompson Flight Company.

Flying Banana (ii)
Reference to the wing shape (apparently influenced by the **Taube**) of the German DFW BI two-seater biplane of the early WW1 period; the essentially similar BII was known at DFW's Leipzig training school as the *Buffalo*.

Flying Banana (iii)
Although *Flying Banana* was suitably

descriptive of the US Navy's Piasecki HRP-1 Rescuer helicopter with its gracefully upswept rear fuselage, it was perhaps used just a little unimaginatively (though just as freely) for the better-known H-21 series (the USAF's Workhorse and the US Army's Shawnee) which had a much sharper bend amidships to provide separation between the tandem rotors.

Banana Boat
Just as there was a degree of rivalry between RAF Lancaster and Halifax crews, so proponents of the B-17 Flying Fortress sometimes used this name as a jibe at the flying boat ancestry of the Consolidated B-24 Liberator (cf *Pregnant Guppy*). Although it was built in greater numbers than any other American warplane in history (18,482) and was in many ways a superior bomber to the B-17, the *Lib* did not always command the same loyalty from its own crews, some of whom dubbed it *Ford's Folly*, from the fact that the Ford's Willow Run, Michigan plant built more B-24s than anyone else. *Flying Boxcar* derived from the deep, capacious fuselage, and is believed to have originated in a newspaper cartoon.

The formal name of Liberator was selected in a company competition, having been submitted anonymously (?) by the wife of founder Major Reuben Fleet, but a dispute arose in April 1942 (after Fleet had sold out) when a faction within the Publicity Dept mounted an unsuccessful campaign to have the name changed to Eagle. Liberator Express was the name given to the 25-seat C-87 transport model, and is not to be confused with the unsuccessful Model 39 Liberator-Liner with a completely new circular-section fuselage and

single-fin tail unit. C-109 was the designation of a fuel-carrying conversion, but a series of take-off accidents earned it the unfortunate epithet of *C-One-Oh-Boom*. The US Navy's PB4Y-2, with a 7ft fuselage stretch and distinctive single-fin, was briefly known as the Sea Liberator before the more familiar title of Privateer was adopted; 26 of the RY-3 transport variant went to the RAF as Liberator C IXs. Privateers flying Firefly missions in Korea, dropping parachute flares in support of night operations, were known colloquially as *Lamp Lighters*.

Banana Bomber
Manufacturers sometimes hold competitions amongst their employees to select a name for a new aircraft, and if one bright spark had had his way in 1960 the **Buccaneer** would have become the Arna, denoting the fact that it was to be A Royal Navy Aircraft; which sounds fine until you remember that Blackburn was the company name which would have gone in front of it. A favourite term of endearment with RAF Lightning and Phantom crews, the name stayed with the *Bucc* throughout a long and often undervalued career, variations on the theme including '*Nana* and the RN's *Peeled Bananas* in overall anti-flash white. **Buccaneer** crews themselves usually preferred *Brick*, which may refer not only to certain flying characteristics but also to the fact that it was built like the proverbial brick outhouse for its punishing low-level strike role; prowess at low level also earned the title *Easy Rider*, and the American term *Dirt Eater* stemmed from 208 Squadron's RAF debut with the type at the 1977 Red Flag exercises.

11

Bandit

Affectionate nickname among some UK operators of the Embraer EMB-110 Bandeirante (**Pioneer**) commuter aircraft. The EMB-111A Patrulha maritime variant is known in its native Brazil by the hybrid nickname *Bandeirulha*.

Banjo

Popular name in the US Navy for the F2H Banshee, the jet fighter that really established the McDonnell Aircraft Corporation, which was still only seven years old when the first of 895 F2Hs flew in 1947.

Bantam Bomber

The Douglas A-4 Skyhawk may have lacked some of the glamour of other US Navy jets but as an example of the designer's art it has few peers, and within days of its first flight in June 1954 the Los Angeles Press had christened the compact little attack aircraft the *Bantam Bomber* and *Heinemann*'s *Hot Rod* after 'Mr Naval Aviation', the great Ed Heinemann. In the Navy it attracted the traditional small-plane names of *Tinker Toy*, *Mighty Mite*, and (more commonly) *Scooter*, whilst the A-4F and subsequent models with the dorsal avionics 'hump' naturally became *Camels*; *Ford* arose out of the pre-1962 designation of A4D and is more closely associated with the F4D Skyray, and *Skyhog* is of course a play on the formal name of Skyhawk.

For an attack aircraft, the A-4 proved remarkably agile and as well as becoming the only non-fighter type to fly with the Blue Angels flight demonstration team it was also adapted as the *Mongoose* aggressor aircraft by the Navy Fighter Weapons School, better known as Top Gun. The NFWS also operates the *Super*

Fox, converted from the standard A-4F by removing the dorsal hump and wing-root guns, and installing a more powerful J52-P-408A engine which gives it a genuinely fighter-style thrust-to-weight ratio of nearly one-to-one.

In the late eighties the Royal New Zealand Air Force updated twenty of its A-4K (for Kiwi?) and TA-4K *Squawks* to Kahu standard, with a new APG-65 radar, HOTAS (Hands On Throttle And Stick) controls, and compatibility with Maverick and Sidewinder missiles; it is claimed the whole package gave the Kahu (the Maori for 'Hawk') 90 per cent of the capability of the Lockheed F-16 at a quarter of the cost, although of course the Kahu remains subsonic like all A-4s. Singapore's A-4S-1 Super Skyhawk also features updated avionics, plus a completely new non-afterburning GE F404 engine. In Israel the A-4 has the alternative name of Ahit (Vulture) and Royal Australian Navy A-4Gs were known to the Air Force as *Chickenhawks*.

Barge

The US Navy's faithful Douglas SBD Dauntless dive-bomber seldom enjoyed the recognition it deserved, and its relative lack of glamour was reflected in such titles as *Barge* and *Clunk*. *Speedy D* and *Speedy 3* (the SBD-3 model) derived from the SBD designation which, paradoxically, was also claimed by its crews to stand for 'Slow but Deadly'. In fact the Dauntless was no mere bomb truck, and was credited with the destruction of nearly 140 aircraft in air-to-air combat. The USAAF's relatively unsuccessful A-24 variant was at one stage named Banshee.

Barling Bomber
After the disaster of the Tarrant
Tabor crash at Farnborough in May
1919, designer Walter Barling took his
ideas to the USA, and under the
patronage of Billy Mitchell had a
single example of the superficially
similar XNBL-1 built by Witteman-
Lewis of New Jersey in 1923.
Variously dubbed the *Barling Bomber*
and *Mitchell's Folly*, the 120ft-span
triplane proved hopelessly
underpowered and was broken up in
1928.

Flying Barndoor
Reference to the thick, broad wing of
the Armstrong Whitworth Whitley,
the incidence of which also gave the
twin-engined bomber its peculiar
nose-down attitude in level flight. As
a parachute trainer at Ringway
(Manchester) the type became
notorious for the 'Whitley Kiss' as
untold numbers of trainees struck
their faces as they dropped through
the aperture in the fuselage floor, and
the Mk 5 glider tug, although never
used operationally, was sometimes
referred to as the *Wombat* within the
airborne forces.

Barra
Usual nickname for the Royal Navy's
fearsome looking Fairey Barracuda
torpedo bomber of WW2.
Barraweewee has also been noted,
although the exact origin appears to
be lost.

Barrel
Though it is hardly the most flattering
description, the 1948-vintage Saab J-
29 swept-wing jet fighter became so
widely known by the nickname
Tunnan (*Barrel*) that it is now
frequently quoted as its formal title.

Flying Barrel (i)
Whatever other qualities they may
have had, Grumman's piston fighters
were not generally noted for their
beauty. The trend began in 1931 with
the original FF-1 *Fifi* two-seat biplane,
also known as the *Fertile Myrtle*
(American pronunciation necessary)
because of its large belly and
retractable undercarriage. Canadian
Car & Foundry licence-built the G-23
derivative as the Goblin, and 40 of
these found their way via Turkey to
the Spanish Republicans, by whom
they were given the somewhat
euphemistic title *Delfine*; the
opposing Nationalists knew it by the
derisory name of *Pedro Rico* (*Rich
Peter*) after its fat and therefore
'affluent' appearance. Although the
term *Flying Barrel* is often used
indiscriminately for all the Grumman
biplane fighters it really belongs to the
even more corpulent F2F and F3F
single-seaters, the FF-1 being so well-
known at the time as the *Fifi*.

Flying Barrel (ii)
Not for nothing were Brewster aircraft
known as *Poor Man's Grummans*,
and even some of the nicknames for
their much-maligned F2A Buffalo
fighter were shared with their Long
Island (NY) neighbour's types,
notably *Flying Barrel* and *Peanut
Special*, although the British variation
Beer Barrel appears unique to the
F2A. Historically important as the US
Navy's first monoplane fighter and
remembered in Britain (where it was
first named Buffalo) for its disastrous
showing with the RAF against Japan's
Zero, the Buffalo found its niche in
Finland where the B-239 model was
affectionately known as the *Pylly
Walteri* (*Bustling Walter*) and even
Taivaan Helmi (*Sky Pearl*). The
Finnish State Aircraft factory also

flight tested in 1944 a local derivative with a wooden wing and Soviet M-63 radial engine, known as the Humu (the closest translation of which is 'reckless').

Basset
RAF version of the racy-looking Beagle 206 five-to eight-seat twin. A total of twenty Basset CC1s were delivered to the Northern, Southern & Metropolitan Communications squadrons (later 26, 207 and 32 Squadrons) from 1965, and used largely for the positioning of **V-Bomber** crews. One Basset (XS770) was attached to the Queen's Flight for Prince Charles's twin-engine conversion training, becoming known as the *Regal Beagle*.

Bat Bomber
Reputedly the most expensive aircraft ever built (the whole programme was capped at $44.5bn in return for just twenty aircraft) the Northrop B-2 Stealth Bomber entered USAF service in December 1993 without a formal name, and was not officially christened **Spirit** until the following spring. Nicknames are relatively scarce as yet but include *Bat Bomber* and *Strategic Boomerang* – the latter referring of course primarily to the flying wing configuration that is something of a Northrop trademark – and is perhaps best remembered for the XB-35 and YB-49 bomber prototypes of the late forties. In a curious reprise of the Rockwell B-1 Excalibur affair, Northrop became involved in legal action in 1990 against a Los Angeles company marketing 'Stealth' brand condoms packaged in the shape of the B-2.

Flying Bathtub
Also known as the *Airknocker*

because of the sound from its tiny two-cylinder engine (typically 26hp) the Aeronca C-2 single-seater monoplane with its deep and narrow fuselage was one of the more distinctive pre-war lightplanes. The C-3 Collegian two-seater – or Master without the 'razorback' look to the fuselage – was also built under licence by the Aeronautical Corporation of Great Britain, and known as the Aeronca 100 when fitted with a 40hp JAP J-99 (an Aeronca E 113C engine built in the UK by J.A. Prestwich). In contrast to the success of the Aeroncas in their American homeland, the British company only succeeded in selling about 24 aircraft, including two with slightly wider fuselages under the new name of Ely 700.

Bayonne
Built by the Groupe Technique de Cannes in the occupied French Riviera, the prototype of the SNCASO SO-90 series of light transports had one of aviation's more dramatic maiden flights. With permission granted only for taxiing trials on 16 August 1943, the Italian troops guarding the airfield were duped into removing certain barbed-wire fences and with nine people on board the twin-engined machine promptly took off on a three-hour flight to freedom in Algeria. Retrospectively named **Bayonne**, the original SO-90 was returned to France after the liberation, but attempts to market a commercial variant (at one stage to have been named Cassiopée) were disappointing: all fifteen of the SO-94 model and the vast majority of the 45 enlarged SO-95 Corse IIs (Corsica) were off-loaded onto the Aeronavale as trainers and liaison aircraft (cf **Bellatrix**).

Beast (i)
RAF nickname for the Vickers Vildebeest torpedo bomber biplane, the formal name of which was spelt with a final 'e' (i.e. Vildebeeste) until 1934. The general-purpose Vincent was closely related, but could usually be distinguished by an external fuel tank between the undercarriage legs.

Beast (ii)
The US Navy's Curtiss SB2C Helldiver was arguably the most despised American warplane of WW2. Someone once said it had 'more bugs than an Oriental flophouse' (between prototype and production stages there were a staggering 889 major modifications to the design) and maintenance men used to say that the SB2C designation really meant *Sonofabitch, 2nd Class*. Nevertheless, the *Beast* as it became most widely known, made a major contribution to the US Navy's war in the Pacific, and also served the USAAF in relatively small numbers as the A-25 Shrike (the traditional but hitherto unofficial USAAF name for Curtiss attack types) before reverting to the Navy title of Helldiver (cf **Cleveland**).

Beaufreighter
Unofficial name of the Bristol Beaufort IX, an Australian freighter/passenger conversion of the locally-built Mk VIII torpedo bomber with a streamlined fairing in place of the dorsal turret. At least 48 conversions were carried out in the closing stages of WW2.

Bébé
Although the fond nickname of *Bébé* is sometimes applied indiscriminately to some of its family descendants, it really belongs solely to the French Nieuport 11 scout of 1915.

Bedcheck Charlie
Whilst the American expression '*Bedcheck Charlie*' dates back to at least WW2 and Guadalcanal, it is perhaps most closely associated with Communist aircraft making night-time nuisance raids in Korea. Usually flown by Yak-18s (NATO – Max) or Po-2s (NATO – Mule), the raids did relatively little material damage, but the effect on morale was compounded by the difficulties experienced in engaging such slow-moving machines at night.

Bedford
Name originally proposed by Bristol for the RAF's Bombay twin-engined bomber-transport, but rejected by the Air Ministry in February 1937.

Flying Bedstead
Floating serenely above the ground to a backdrop of Pathe words and music, its pilot perched precariously on top, the aptly named *Flying Bedstead* became one of the abiding images of fifties Britain. Officially the Rolls-Royce Thrust Measuring Rig (TMR) and powered by a back-to-back pair of Nene turbojets, two machines (XJ314 and the short-lived XK426) were built for VTOL research. The nickname was later purloined for NASA's Bell LLRVs (Lunar Landing Research Vehicle), one of which crashed at Ellington AFB in 1968, piloted by a certain Neil Armstrong.

Beechjet 400
New name adopted by Beech for the Mitsubishi Mu-300 Diamond II corporate jet, after acquiring the rights from the Japanese company in December 1985; prior to the deal Mitsubishi had delivered 100 Diamond Is and IIs in kit form for assembly by its American branch at San Angelo,

Texas, and the first 65 Beechjets also included Japanese components. In February 1990 the Beechjet 400T won the USAF's TTTS contest (Tanker/ Transport Training System) and the first T-1A Jayhawk (a name clashing with that of the Coast Guard's HH-60J helicopter – cf *Catfish*) was delivered in January 1992.

Beercat
Armée de l'Air name for the Grumman F8F Bearcat fighter, used in French operations in Indochina between 1951 and 1956. The Bearcat was Grumman's last single-engined piston fighter, and came just too late for US Navy service in WW2.

Flying Beermat
In many ways analogous with the better-known Vought *Flying Pancake*, this intriguing circular-winged, single-engined machine was apparently designed and built by an unidentified German farmer near Leipzig in 1944. Details are sketchy, but it is thought the amateur fighter project was abandoned unflown after the undercarriage was damaged in taxiing trials.

Flying Beetle
Popular term for the Shorts SC1 VTOL research aircraft, with its curious appearance of broad delta wings, round dumpy fuselage, and long fixed undercarriage. The SC1 was powered by five Rolls-Royce RB 108 turbojets (four for lift and one for forward propulsion) and the first of two had its maiden flight on 2 April 1957, making it Britain's first fixed-wing VTOL type.

Beetle in Boots
Nickname for the prototype of the diminutive Percival Mew Gull racer

(G-ACND) of 1934, initially fitted with an oversize, spatted undercarriage similar to that of the larger Gull series 1 and 2 tourers.

Bellatrix
As well as the **Bayonne**, the remarkable Group Technique de Cannes also built a prototype 23-seat pressurised airliner, the SO-30N **Bellatrix**, which was dismantled and secretly stored until its maiden flight in February 1945. A pair of tricycle-geared SO-30R development aircraft bore the same name, but this was dropped in favour of Bretagne (Brittany) for the 45 SO-30P production airliners.

Bentwing Monster
With its distinctive cranked wing and powerful radial engine, the US Navy's WW2 Chance Vought F4U Corsair fighter soon picked up a string of nicknames like *Bentwing Monster* (or *-Bird*), *Horseshoe*, *Super Stuka*, *U-Bird* (perhaps also connected to the designation), *Hose Nose* and *Hog Nose*, and there were even jokes about seagulls learning to scare Japanese pilots by flying upside-down. Early models with the framed canopy were known as *Birdcage Corsairs*, and the Japanese supposedly knew the F4U as the *Whistling Death* after the sound made by the leading-edge intakes, although this is now acknowledged as apocryphal. The Corsair was not without its vices though, and an early reputation for poor carrier landings earned it the familiar US Navy epithet of *Ensign-Eater* (or *-Eliminator*) but once the wrinkles had been ironed out the F4U went from strength to strength, hitting the headlines in 1945 as the *Sweetheart* (or *-Angel*) of Okinawa, and seeing action as late as 1972 when

El Salvador used its ancient Goodyear-built FGs against an uprising.

Berg
Alternative name for both the Austrian Aviatik DI scout and generally similar CI recce biplane of 1917, after their designer, Julius von Berg.

Biffo
One of the real workhorses of post-war British aviation was the simple and effective Bristol 170, known in passenger carrying configuration (with secondary freight capability) as the Wayfarer, and in more familiar nose-loading cargo form as the Freighter; the Mk 21 series with extended round-tipped wings and more powerful Hercules 672 engines was initially billed as the New Freighter. In everyday use 'Bristol Freighter' tended to cover all types, and in turn led to the popular nickname of *Biffo* after the contemporary 'Biffo the Bear' comic character. Easily the most famous of all Freighter operations was Silver City's cross-channel car ferry service out of Lympne and later from the purpose-built Ferryfield (Lydd), which is now looked back upon with such nostalgia. The service began in July 1948 with a Mk 1 Freighter, and later the Mk 32 was developed specifically for the airline with an enlarged tail and lengthened nose to provide accommodation for three medium-sized cars with 23 passengers in seats at the rear. Silver City called the Mk 32 the Superfreighter or Super Bristol, and in 1958 converted the first (G-AMWA) into 60-seat, all-passenger configuration as the Super Wayfarer. Less successful was Jersey Airlines' operation of a single Mk 31E Baronet

class car ferry (G-AMLL) between Jersey and Dinard (Brittany) in 1957/58, and BEA also had just a single Mk 21 Yeoman class aircraft (G-AICS) used only intermittently as a general-purpose freighter before being leased to Silver City in June 1952. Amongst other operators were the RCAF, RAAF, RNZAF, Pakistan Air Force, Aer Lingus, Air Charter, Sabena and Trans Canada, the latter allocating the name Cargo Liner.

Big Ack
With its maker's initials heavily embossed on the nose, the RFC's Armstrong Whitworth FK3 recce biplane (resembling the BE2) was widely known as the *Ack W* in line with the phonetic alphabet of the day; when the larger FK8 appeared at the Front early in 1917, again with the AW initials on the nose, the two types became differentiated by the nicknames *Little Ack* (FK3) and *Big Ack* (FK8).

Big Ant
Occasional nickname in the West for the Antonov An-124 Ruslan (NATO – Condor) heavy freighter, and ironically shared with the An-2 biplane (cf *Little Annie*).

Big Bird
Original name of the Quickie Aircraft Corp. Free Enterprise, purpose-built for an attempt at the first non-stop, unrefuelled circumnavigation of the world. The flight was envisaged as taking about 90 hours, using an autopilot to allow the single crewman to sleep, but sadly the Free Enterprise crashed at Mojave, California on 2 July 1982, killing company founder Tom Jewett. Ironically the crash was witnessed by Jeana Yeager and Dick Rutan, who were already planning the

remarkable Voyager in which they were ultimately to accomplish aviation's last big 'first' between 14 and 23 December 1986.

Billy's Bomber
Occasional term for the WW2 North American B-25 Mitchell light bomber, named after Brigadier General William 'Billy' Mitchell, who did so much to establish US air power.

Bird Cage
Common term among pre-WW1 aviators for the famous Blériot monoplanes with their uncovered, wire-braced rear fuselages. The name derived from the old chestnut about a French rigger explaining to an upper-class English flier that the only way to check the mass of wires was to release a bird into the fuselage each morning and see if it escaped.

Bitch
Heartfelt epithet at Canadian Car & Foundry (CCF) for the troublesome CBY-3 Loadmaster 'lifting-fuselage' transport (one only, CF-BEL) built in 1945 to the designs of Burnelli Aircraft of the USA. An earlier manifestation of Vincent Burnelli's concept for aircraft with broad aerofoil-section fuselages, the UB-14, was to have been built in the UK as the RR Kestrel-powered Clyde Clipper, until the specially formed Scottish Aircraft & Engineering Co. went into receivership in early 1937. A British redesign did eventually fly as the Cunliffe-Owen OA-1 (G-AFMB) however, only to meet an ignominious end on a VJ night bonfire in Egypt.

Blackbird
In America and Europe, *Blackbird* became so firmly established as a title for the exotic Mach-3 Lockheed SR-

71 strategic recce aircraft that its unoffical status may seem somewhat irrelevant, but in fact in other parts of the world it became just as well-known by the title of *Habu* (a venomous Japanese snake) acquired during its first deployment to Kadena in 1968. Also known to its crews as the *Sled* (though hardly a *Lead Sled*), the SR-71 actually inherited the famous *Blackbird* name from the slightly smaller and faster YF-12 experimental interceptor, which with its advanced weapons system and AIM-47A Super Falcon missiles seemed even more like something out of 'Dan Dare'. The YF-12s in turn were adaptations of the original A-12 reconnaissance aircraft (the bizarrely named Oxcart programme) known to its crews as the *Cygnus*, in keeping with Lockheed's series of 'Heavenly Body' names. The A-12 was first revealed by President Johnson in early 1964 when he mistakenly (?) referred to it as the 'A-11', and later the same year LBJ supposedly repeated his blunder by muddling the correct RS-71 designation of its successor; the popular story is that the 'SR-71' version was allowed to stand simply because no one dared correct the President of the United States, but it seems more likely that the whole episode was part of the deliberate obfuscation that accompanies most 'Black' programmes in the US.

Black Bomber
Widely used but unofficial title of the 1953 Vickers Valiant B2, a beefed-up, low-level pathfinder variant of the standard Valiant B1 distinguished by its gloss black colour scheme and its pods housing the main gear projecting from the trailing-edge of the wing. Unlike its counterparts the Vulcan and Victor, the Mk 2 version of the

Valiant was to progress no further than a single prototype (WJ954) and in early 1965, less than a year after the **V-Bomber** force switched to low-level operations, the RAF's Valiant B1 fleet was to be permanently grounded with fatigue problems.

Black Bullet
Slightly curious nickname for the magnesium Northrop XP-56 pusher-engined fighter, given that the only black paint was on the spinner of the second and final machine.

Blackhawk
The Curtiss proposal of **Blackhawk** as a name for the four-jet XF-87 night-fighter was supposedly rejected by the USAF on the grounds that the hawk is a diurnal creature, and the occasional reference to the name *Nighthawk* in connection with the XF-87 may have been an attempt to overcome this objection whilst maintaining the company tradition of Hawk titles. Sadly, that tradition and the company itself came to a disappointing end when the overweight and underpowered XF-87 was cancelled in October 1948 after just a single prototype.

Black Manta
In some respects the most mysterious of all American 'Black' (i.e. classified) programmes of recent years is the supposed Northrop TR-3A **Black Manta** reconnaissance aircraft, said to have evolved from the company's unsuccessful design for the XST programme (cf *Wobblin' Goblin*). Since an initial flurry of reports in 1990 and 1991 little has been heard of the TR-3A, so both the validity of the name and the existence of the aircraft itself must remain open to question.

Black Widow II
With its deadly, stealthy appearance and its pedigree, there should never have been any doubt about a name for the Northrop/McDonnell Douglas YF-23A fighter. Northrop employees made the point by surreptitiously doubling up the small red triangle calibration mark under one of the prototypes to form a pair of sinister-looking eyes, as a result of which officialdom had a nasty attack of arachnophobia. Northrop personnel supporting the evaluation at Edwards AFB were expressly forbidden to use the name, although one hapless company official in Washington had to confess to an aviation journalist that he didn't even know his company had built the original P-61 Black Widow night-fighter in WW2, let alone that there was any attempt to resurrect the name. The pair of YF-23s were also sometimes known by the *Grey Ghost* callsign used during the evaluation, but the whole question of a name became academic on 23 April 1991, when the USAF announced the rival Lockheed/Boeing/General Dynamics YF-22A as winner of the Advanced Tactical Fighter (ATF) competition.

Black Widow-Maker
Fortunately, the Boulton Paul P120 research delta failed to live up to its gruesome nickname when tailplane flutter caused it to crash near Boscombe Down on 28 August 1952 just three weeks after its troublesome maiden flight; chief test pilot Ben Gunn escaped virtually unhurt, despite ejecting inverted at no more than 3,000ft.

Blarebok
The South African Air Force's AM-3C Bosbok (Bushbuck) forward air control aircraft, assembled locally by

Atlas from Aermacchi components, was once described beauitifully by a *Flight* man as making a sound on take-off 'like a Shackleton falling downstairs'. From the Bosbok, Atlas also developed the utility C4M Kudu (a species of antelope), 'a strange and terrible device' with a capacity for turning fuel into noise that earned it the title of the *Converter*. Both the Bosbok and the Kudu are descendants of the 1959 Lockheed Georgia LASA-60 high-winged utility aircraft, which was built briefly in Mexico by Lockheed-Azcarate as the Aeroguayin, and more successfully by Aermacchi as the AL60 Santa Maria. The Italian company also developed the AL60-F5 variant, ten of which beat sanctions to serve with the Rhodesian Air Force as the Trojan, and the AL60-C5 Conestoga, which first introduced the taildragger layout of the Bosbok and Kudu and was sold to the Central African Republic. The slimline AM-3, eventually to mature as the Bosbok, appeared in 1967.

Blaster
Blaster was the name favoured by Brewster Aeronautical for the woefully inadequate SB2A **Buccaneer** scout bomber it produced for the US Navy in WW2. Just over 200 are believed to have been delivered to the RAF as the Bermuda, and after being euphemistically deemed 'unsuitable' for operational use most were simply scrapped.

Blenburgher
Early nickname for the much-maligned Bristol Blenheim light bomber. The Blenheim IV with a new extended assymetrical nose in place of the original short unit (inherited from the commercial Type 142 'Britain First') was initially to have been

known as the Bolingbroke; the name was dropped in the UK, but was used for the 676 aircraft built by Fairchild Aircraft Ltd in Canada, also sometimes known by the diminutive *Boly*. An armoured close-support prototype with a 'solid' four-gun nose (AD657) was appropriately known as the Bisley after the Surrey home of the National Rifle Association, but although the name was dropped in favour of Blenheim V for the developed bomber version with a half-glazed nose, it stayed informally with the Mk V throughout its RAF service. One example of the unpopular Mk V was stripped of its armour and turret in the Middle East and used as a high-speed hack with the informal name of *Wisley*.

Blitz
Blitz (*Lightning*) was the widely used and wholly appropriate nickname for the influential (just look at the Spitfire) Heinkel He 70 high-speed, single-engined transport.

Bloater
Although **Bloater** was to become synonymous with the RFC's unloved Royal Aircraft Factory BE8 two-seater, in his autobiography *Flying Fury* WW1 ace James McCudden also uses it for its forerunner, the BE4. The exact origins of the name are uncertain, but it could possibly be a play on 'B.E. Rotary' to distinguish it from the better known BE2 series with inline motors (cf *Bognor Bloater*).

Blockbuster
Sardonic term for the WW1 Sopwith 5F1 Dolphin fighter, with the pilot's head projecting through the upper wing and thus vulnerable to decapitation in the event of the aircraft overturning on landing.

Bloody Paralyser
Occasional reference to the Handley
Page 0/100 and 0/400 heavy bombers,
after a plea by Cdr Charles Samson
during the German advance on
Antwerp in 1914 for 'a bloody
paralyser of an aeroplane to stop the
Hun in his tracks'. **Handley Page** types
were also known phonetically in WW1
as *H-Pips*, and the four-engined 126ft-
span V/1500 was sometimes referred
to as the *Super Handley* to distinguish
it from the 0/100 and 0/400 twins.

Bludgeon
Apt name originally proposed by
Bristol for their 1927 Type 95 Bagshot
(J7767), an experimental twin-engined
night-fighter monoplane fitted with a
pair of 37mm Coventry Ordnance
Works weapons known colloquially as
'Cow Guns'.

Blue Canoe
Popular USAF term for the Cessna L-
27 liaison aircraft, a militarised
version of the successful civil Model
310 light twin; a total of 195 were
procured between 1956 and 1961, and
re-designated as U-3s in 1962.
Colemill Enterprises of Nashville offer
a 310 (Models F to Q) upgrade known
as the Executive 600, with new 300hp
Continental IO-520 flat-sixes and
three-blade McCauley props.

Blue Circle Fighter
Development problems with the
Marconi (GEC) Foxhunter radar in
the late eighties meant that an
embarrassing number of RAF
Tornado F2 and F3 interceptors were
forced to fly with concrete ballast in
their noses, leading to this scathing
nickname bestowed in honour of the
Blue Circle cement company.
Otherwise, the variable-geometry
Panavia Tornado has been known in

the RAF as the *Swinger*, with the
ground attack GR1 and air defence F3
also being the *Can Opener* and *Flick
Knife* respectively; Italian Tornadoes
in the Gulf War were known
colloquially as *Locusts* after the
Operation Locust name used for the
deployment.
Panther was the provisional name
used in the early days of the Anglo/
German/Italian programme, with the
Panther 200 eventually maturing into
today's Tornado IDS (Interdictor
Strike) and Panther 100 referring to a
projected single-seat Luftwaffe
model. The Panther name was little
used however, and until the Tornado
title was formally adopted on 10
March 1976 it was usually known
simply as the MRCA, officially
signifying 'Multi-Role Combat
Aircraft', but to the doubters 'Mother
Riley's Cardboard Aircraft' or 'Must
Refurbish Canberra Again'.

Blunty
Affectionate term within the Royal
New Zealand Air Force (RNZAF) for
the BAC/BAe Strikemaster Mk 88.
The Strikemaster was of course an
armed export derivative of the RAF's
long-serving Jet Provost trainer (often
known simply as the JP) but almost
forgotten now is the Eaglet proposal
to the USAF as a Cessna T-37
replacement in the seventies, based on
the Strikemaster airframe but
powered by a Garrett TFE-731.

Bobsleigh
First flown in July 1945, the sole Reid
& Sigrist RS3 Desford twin-engined
trainer (G-AGOS) was purchased by
the Air Council four years later and
modified with an elongated glazed
nose for prone-pilot research as the
RS4 **Bobsleigh** (VZ728). In 1956 it
was restored to the civil register as the

RS3 Desford Trainer, and enjoyed a long career as a survey and airborne photography aircraft.

Bognor Bloater
Not to be confused with the better-known BE8 *Bloater*, the White & Thompson two-seater biplane (sometimes referred to as the Type 1172) earned its memorable nickname from its birthplace and from the 'scaly' effect created by the copper wire stitching on its wooden monocoque fuselage. Ten were built for the RNAS from 1915.

Boiler
An early and important essay in cabin pressurisation, the Lockheed XC-35 was based on the commercial Model 10 Electra, and earned its nickname when faulty door seals produced a disconcerting hiss the first time pressure was applied.

Bolshie
Typical of Miles' enterprise, the LR5 (M64) was a spare-time effort by a group of employees to develop a two-seater club aircraft for the post-war market, abandoned when flight testing in June 1945 revealed unacceptable take-off and landing characteristics.

Boomer
In Australia a boomer is a large male kangaroo, so it was perhaps a doubly appropriate nickname for the Commonwealth CA-12/13/19 **Boomerang**, given its reputation for very low-level pop-up attacks in support of Australian forces in New Guinea, Borneo and the Solomon Islands. Also known as the *Diggers' Delight*, the **Boomerang** was based largely on the Wirraway trainer (cf *Pilot Maker*) and first flew at Fisherman's Bend in May 1942.

Boomerang
An unidentified Australian pilot in RAF Coastal Command bestowed this title on the Lockheed Hudson, in recognition of the way it could be relied upon to come back after a mission. The exploits of the often overlooked Hudson included the RAF's first UK-based kill of WW2 (a Do 18), taking part in the first thousand-bomber raids, landing Agents ('Joes') in occupied France, and even capturing a U-boat intact. BOAC also operated Loch class Hudson IIIs on wartime Scrutator flights to Sweden. Originally a highly controversial purchase (along with the NA Harvard) for the RAF, the Hudson was a derivative of the commerical Model 14 Super Electra, a type known to Northwest Airlines as the Sky Zephyr. The Rausch Aviation *Hudstar* was a one-off hybrid fitted with the tail and rear fuselage of the stretched Model 18 Lodestar.

Boston
Britain first began receiving Douglas DB-7 light-bombers as Bostons in mid-1940, but the first variant to see active service with the RAF, the following spring, was the Havoc night-fighter. Modified for the role by the Burtonwood Repair Depot, the Havoc I was a converted Boston II with AI radar and eight 0.303 Brownings in the nose, whilst the later Havoc II had twelve guns. Moonfighter, **Ranger** and Havoc IV were all names applied successively to a special fighter-bomber version (without radar) used by 23 Squadron for Intruder sorties over France. Two unusual night-fighter experiments were the unarmed Helmore (later Turbinlite Havoc) with radar and a 2,700 million candlepower GEC searchlight to illuminate enemy

bombers for accompanying Hurricanes, and the Pandora, which was equipped in the bomb bay with Long Aerial Mines attached to a 2,000ft cable and dropped into the path of oncoming bomber formations; conversion totals were 20 and 70 respectively, but neither scheme proved really successful. The Havoc name was later taken up by the USAAF for its mass-produced A-20 attack model, although the RAF kept the name Boston for its light-bombers. Nighthawk was the seldom-used name for the American P-70 night-fighter, which saw strictly limited service before giving way to the Northrop P-61 Black Widow.

Bother
Suitably apt pun on the Blackburn Botha (properly pronounced 'Boather') for the distinctly lacklustre torpedo bomber which saw limited service with the RAF early in WW2.

Flying Boudoir
Conceived in the days before synchronised guns firing through the propeller arc, the Sage Type 2 had a gun in a cupola on the upper wing and the unprecedented luxury of an enclosed cabin to shield the observer from the slipstream as he stood up to fire it. By the time the sole Type 2 flew in August 1916 synchronisation gear had become available, and the project was abandoned following its crash near Cranwell a few weeks later.

Box
Along with their 6ft 6in square fuselage cross-section, the Shorts 330 (ex-SD3-30) and 360 commuter airliner also inherited the affectionate names of *Box*, *Shoebox* and *Super Shed* from their predecessor the Skyvan. The SD3-MR Seeker maritime variant of

the 330 failed to sell into a market already glutted with such types when it was announced in late 1977, but Shorts had more success with the Sherpa freighter featuring a Skyvan type rear-loading ramp, achieving notable sales in the late eighties to the USAF and US Army as the C-23. The 330-UTT (Utility Tactical Transport) which sold to Thailand is similar, but lacks the Sherpa's rear ramp and strengthened floor. Philippine Airlines uses the name Sunriser for the Shorts 360, the 330's bigger, single-finned brother.

Flying Boxcar
Soviet term for the Yak-24 (NATO – Horse) tandem-rotor helicopter, with its long, slab-sided fuselage, first flown in 1953.

Flying Brickyard
Early reference to the 34,000 heat resistant tiles (no two of which are identical) used to protect the Space Shuttle Orbiter during re-entry into the Earth's atmosphere.

Brimstone Bomber
Journalese term for firefighting or water-bombing aircraft, usually conversions of existing types. The only purpose-designed firefighter to enter service so far (in 1969) is the Canadair CL-215 (CL-415 in turboprop form) amphibian, but the most impressive firefighters of all must be the pair of 200ft-span, late-war-vintage Martin Mars flying boats operated in Canada by Forest Industries Flying Tanker.

Brisfit
That simple yet remarkably effective name 'Bristol Fighter' merely goes to underline the impact made by the Bristol F2A/F2B two-seater. The *Biff*, as it was better known in WW1, took

a mauling from Richthofen's Albatrosses on its very first outing in 'Bloody April' 1917, but when it was properly employed as a scout with a sting in the tail, it gained a fearsome reputation in the German Jagdstaffeln. The last Bristol Fighter was built in 1927, and the type remained in service with the RAF as an army cooperation machine until as late as 1932.

A single civilianised Bristol Type 27 Coupé conversion appeared in 1919 with a hinged cover over the rear cockpit, and it was followed by 28 Type 28, 29, 47 and 48 Tourers, most of which had a widened rear fuselage to allow seating for two in the former gunner's position. A more ambitious derivative was the one-off Type 36 Seely (G-EAUE) two-seater of 1919, with a deeper fuselage, eliminating the usual gap between fuselage and lower wing.

Britannic
Title chosen by Shorts in 1958 for what became the RAF's Belfast strategic freighter, reflecting the fact that it utilised the flying surfaces of the Bristol Britannia, 30 of which were also built by Shorts. Suction drag on the upswept rear fuselage earned the Belfast the title of *Dragmaster* during trials, but by modifying the tailcone and removing the yaw-damping strakes it became the *Fastback*. An intriguing 1962 proposal to marry the capacious Belfast fuselage to the swept wings of the Lockheed (Georgia) C-141 Starlifter (cf **Super Hercules**) went under the picturesque but unofficial name of *Georgia Bel*; sadly the project was scuppered when Shorts' management prematurely leaked the story to the British press.

Buccaneer
Bristol name for a 1942 torpedo

fighter project that was eventually to mature as the Brigand twin-engined light-bomber of the post-war RAF.

Buckaroo
The original **Buckaroo** was a tandem two-seat trainer derivative of the Globe GC1 Swift lightplane, developed by Temco after building several hundred Swifts under subcontract and acquiring the rights when Globe went under in 1947. The USAF evaluated the extensively redesigned TE1B version as the YT-35 and the Navy also considered the Model 33 Plebe tricycle-gear development of 1953, but in both instances Temco lost out to the Beech T-34 Mentor. The only customer was Saudi Arabia, who bought ten of the export TE1A model in 1953, although Greece, Israel, Italy and the Philippines all expressed an interest.

The original Globe Swift was at one stage planned for licence production in the UK as the Globe by Helliwells of Walsall, and the sporty appeal of the type also led to attempts to develop updated models in the late eighties, when a pair of elderly Swifts were re-engined by Lo Presti Piper Aircraft Engineering as prototypes for the SwiftFury (180hp Lycoming 0-360 piston engine) and SwiftFire (425hp Allison 250 turboprop). A tricycle-gear version of the former was also proposed for the USAF's enhanced flight screening aircraft programme (eventually won by the Slingsby T-3A Firefly) as the SwiftThunder but the whole Lo Presti programme has been dogged by funding difficulities.

Budgie
Popular name, at least within the aerospace industry, for the workhorse Avro/HS/BAe 748 twin turboprop. The RAF name of Andover covers

both the standard fuselage variant and
the Type 748MF/Type 780 specialised
tactical transport version with
stretched fuselage, rear-loading ramp
and kneeling undercarriage, most of
which were sold off in 1975. An
interesting project was the mid-sixties
Airborne Early Warning (AEW)
Andover with overwing Trent
turbofans (the usual props would have
caused ghosting on the radar) and
front and rear radomes, à la Nimrod
AEW MK 3; the use of two radomes
(and the shape of the front one) gave
rise to the nickname *Toucan* because,
as the designer said, 'two can do the
job better than one can'.

Sadly the AEW version was never
built, but company demonstrator G-
BCDZ appeared in a number of other
guises over the years: as the BAe
Coastguarder it had a brief moment of
glory as the winner of the Sea Search
'81 competition at that year's
Greenham Common IAT but failed to
generate any orders in the
overcrowded light maritime patrol
market, and around the same time it
also demonstrated a 15lb hushkit to
the noisy RR Darts, with which the
type was promoted as the Intercity
748. The one-off Macavia BAe 748
Turbine Tanker (G-BNJK) converted
at Cranfield in 1987 was a firebomber
with a removable 24ft-long belly
pannier capable of dispensing 2,000
US gallons of water. Austrian Airlines
used the class name Belvedere for the
pair of HS 748 Srs 2s it acquired in
1966.

Although the BAe ATP (Advanced
Turboprop) was derived from the
Super 748, component commonality is
said to be limited to about 25 per cent;
upon rationalisation of BAe's
turboprop business in January 1993,
the ATP was renamed Jetstream 61.

BUFF
Supposedly an acronym for Big Ugly
Fat (or Friendly) Fellow (there are
other, less delicate interpretations),
this is a common American term for
aircraft like the Grumman A-6,
Boeing B-52, and Sikorsky CH-53,
among others.

Bug Eye
Popular name for the Optica
observation aircraft with its
helicopter-style 'bubble' cockpit,
originally flown by Edgley Aircraft as
the EA7 in late 1979; the nickname
was particularly prevalent in media
reports of the disastrous fire at Optica
Industries' Old Sarum (Wilts.) factory
in January 1987. Successor company
Brooklands Aircraft renamed it
Optica Scout, and also devised the
Scoutmaster variant with thermal-
imaging and/or search radar, before
the whole project was acquired by
FLS in 1990.

Old Bug Eyes
Pilots' nickname for the USAAF
Douglas C-74 Globemaster I freighter
with its side-by-side fighter-style
canopies. The better known C-124
Globemaster II with a deeper fuselage
and more conventional flight-deck was
known as *Old Shaky* or the
Aluminium Overcast.

Bug Smasher
Military men and heavy metal freaks
are not always the most appreciative
of people when it comes to light
private-owner aircraft, and in
America *Bug Smasher* is a favourite
term for anything small with a fan in
front of it; similarly, *Indians* is
sometimes heard in relation to a
gathering of such aircraft, perhaps
with thoughts of Piper types originally
in mind. The British counterpart is

Spamcan, which probably began as a generic term for Cessnas back in the heady days of the sixties and seventies (before the US Product Liability laws wrecked the industry) when only Dateline had more singles on its books.

Bull

The first 'public appearance' of the grotesque Blackburn Blackburn fleet spotter (named after the Lancashire town) came on 30 September 1922, when the third prototype force-landed in a field near Lowestoft. A few weeks later *Aeroplane* published a photo of the stricken machine along with a comment from C.G. Grey likening it to 'a docile bull in a field', coining a nickname that came to be used for the pair of trainer conversions used by No. 1 FTS, Leuchars.

Bullet

Widely used RFC term for what were then considered high-speed types, such as the Morane type N, Bristol M1 monoplane and Scout biplane, and Vickers ES2 and FB19.

Bullet & Blades

Few would deny that the F-104 Starfighter (or StarFighter, as it was originally styled) was the hottest and sexiest fighter of its generation, and with more than a little help from Lockheed's PR Dept the press were soon gushing over the *Bullet & Blades* (i.e. bullet fuselage and razorblade wings) and the '*Missile with a man in it*'. Despite the hype the early models were of limited practical value, although Pakistani F-104As picked up the grudgingly respectful title of *Badmash* (*Wicked One*) from their Indian opponents in the 1965 conflict, and during two brief and undistinguished tours in Vietnam

USAF F-104Cs became known as *Zippers*.

The F-104 hit the headlines again in the mid-sixties as a result of the appalling (yet hotly debated) loss-rate of West German F-104G Super Starfighters. With an average of a crash every ten days at one stage, epithets like *Widow Maker, Flying Coffin* and *Beautiful Death* proliferated, and so did the black humour; for example it was said that the definition of a true optimist was a Starfighter pilot giving up smoking on health grounds, and that any German wanting his very own F-104 only had to go into his back garden and wait . . .

Japan adopted the local title of Eiko (Glory) for its F-104Js and Italy's advanced F-104S, later the F-104S(ASA), was sometimes referred to as the *Spaghetti Starfighter*, which makes a handy mnemonic, although the designation actually denotes its compatibility with the Sparrow missile. The CL-1200 Lancer was a late-sixties proposal for a development with a new high wing, low-set tail, and either a TF30 or F100 engine.

Bull Pup

'You are old, Mother Chipmunk, the young Pup said'; so ran a poem in the 1969 RAF Souvenir Book (now the Yearbook), and in fact a military variant of the popular Beagle Pup lightplane, the B-121M **Bull Pup**, had been envisaged as long ago as 1965. Sadly though, by the time the project had matured into the B-125 Bulldog (first flown 19 May 1969) the Beagle company was already doomed and the RAF didn't get the new trainer, from Scottish Aviation, until 1973. Almost forgotten now is the one-off Bullfinch tourer (G-BDOG) with an ungainly retractable undercarriage, first flown in 1976 and also projected in military form as the Bulldog 200 series.

Bull's Eye
With financial backing from his brother, whose interests lay in Everard's elastic webbing manufacturers, it was a stock joke among the pre-WW1 aviation fraternity that Alliot Verdon Roe's 'Avroplanes' were held together with Bull's Eye brand trouser braces, hence the nickname applied to the first Roe I triplane of 1909. The second and final Roe I appeared at Brooklands at Easter the following year with the upper two sets of wings extended but the lower unit severely truncated, and was promptly dubbed the *Two-and-a-Bit Plane.*

Bumble Bee (i)
Apt nickname for the stubby little McDonnell XF-85 Goblin experimental parasite fighter of 1948, designed to be carried for self-defence by the Convair B-36 bomber (cf **Peacemaker**).

Bumble Bee (ii)
Project-stage name for the Bristol 171 Sycamore, which on 25 April 1949 became the first British helicopter to be awarded a Certificate of Airworthiness.

Butcher Bird
The German RLM may not have recognised designer Kurt Tank's favoured name of Wurger (Shrike) for his Focke-Wulf Fw 190 fighter, but considering the awesome reputation the 190 attained it was perhaps inevitable that the name should persist in the colloquial form of *Butcher Bird.* The Fw 190D-9, the first production model with the Jumo 213A inline engine (with an annular cowling giving

the appearance of a radial) was quickly named *Dora-9*, but D-series aircraft as a whole were also known as the *Langnasen Dora* or *Long-Nosed Dora*. *Kangaroo* was the predictable nickname of the developmental Fw 190 V18/U1 with its large underslung turbo-supercharger.

Butterfly
Predictable nickname for a single Beech AT-10 Wichita twin-engined trainer fitted in 1944 with an experimental version of the V-tail that was soon to become something of a Beechcraft trademark, appearing on the famous Bonanza and the less well-known Model 34 Twin Quad transport.

Caballero
Name used briefly in 1964 for the Helio H-250A Courier II STOL lightplane. The military U-10 variant was operated by both the USAF and the CIA's 'Air America' in Vietnam where, equipped with loudspeakers and leaflet dispensers for the psychological warfare role, it was known as the *Speaker Bird*.

Cadillac
Nickname of the Northrop M2-F2 lifting-body research aircraft (cf *Flying Stone*) after its fifties auto-style fins.

Camel
Officially, the new Sopwith two-gun fighter was simply the F1 Scout, but coming from the Kingston-upon-Thames 'menagerie', and with that distinctive fuselage 'hump' over the breeches of the Vickers guns, it was always going to be the *Camel*. Having apparently learned nothing from the *Pup* affair, numerous instructions were handed down forbidding use of the name (which naturally had the

opposite effect) and as the sensitive beast went on to establish itself as the most successful fighter of WW1 the Colonel Blimps of the day had little choice but grudgingly to acknowledge the title.

A conversion lacking the famous hump was the successful night-fighter adaptation believed to have been devised by 44 Squadron in late 1917, which inherited the name *Comic* from an earlier 1½ Strutter conversion (cf *Flapping Goose*). Both Vickers guns were removed and the top decking faired over, twin Lewises fitted on the top wing on SE5 mountings with a 30¼ inch cut-out between, and the cockpit moved aft eighteen inches to make way for a BE2e fuel tank.

Canberra
Temporary type name (and applied on an individual basis to the prototype G-EBTL) of the de Havilland DH 61 Giant Moth single-engined transport biplane, designed in 1927 to meet an Australian requirement.

Caravan II
A joint development by Cessna and licensee Reims Aviation, the French-built F406 Caravan II twin is essentially a Cessna 404 Titan with P&WC PT6A-112 turboprops in place of the former Continental piston units. The Caravan II first flew in 1983, and should not be confused with the single-engined Caravan I (cf **Cargomaster**).

Cargo Commuterliner
Title used by Consolidated Aircraft to market the Israel Aircraft Industries 101B Arava twin-boom STOL transport in the USA; first deliveries under the new name were to Key West Airlines in mid-1981. Sherpa was the title used around 1970 for a projected

version of the Arava with Astazou turboprops, to have been built under licence by Socata of France.

Cargomaster
Name used by American overnight package carrier Federal Express (Fed Ex) for the Cessna 208A, a customised all-cargo version of the standard Model 208 Caravan I utility aircraft; Super Cargomaster applies to the stretched Model 208B (Grand Caravan in fourteen-passenger form) and both differ from standard machines in their underbelly panniers and lack of cabin windows. First flown in December 1982, the Caravan I (military U-27) was a wholly new design unrelated to other Cessna singles; nor should it be confused with the Reims-built **Caravan II** twin.

Caribou II
Early title for the de Havilland Canada DHC-5 Buffalo twin-turboprop STOL tactical transport, evolved from the piston-engined DHC-4 Caribou. A 40-seat commercial version was launched in 1979 as the DHC-5E Transporter, but the only customer was Ethiopian Airlines, taking just two.

Caspian Sea Monster
Nickname within the US intelligence community for the intriguing Soviet Wing-In-Ground Effect (WIGE) craft which apparently came to a sticky end one foggy day sometime in the late sixties. WIGEs – or Ekranoplans (Wingships) as they are known in the CIS – 'fly' low over the surface on a trapped cushion of air, and the KM version in question allegedly had a length of over 100 metres, was powered by ten jet engines, and once left the water at a weight of no less than 540 metric tons. Other WIGEs

have come and gone since, in the CIS and elsewhere, but there will probably never be another to compare with the mysterious *Caspian Sea Monster*.

Cat

Despite the fact that it was already approaching obsolescence by 1939, the twin-engined Consolidated PBY went on to become arguably the world's most famous flying boat during the course of WW2, and certainly the most prolific. The US only adopted the British title of Catalina, suggested by Consolidated founder Reuben Fleet, in October 1941, and subsequent role-specific nicknames included the *Mad Cat* with early Magnetic Airborne Detector equipment, the famous nocturnal *Black Cat*, and the *Dumbo* rescue aircraft. *Pig Boat* was a corruption of *P-Boat* (from the PBY designation) and Dutch Navy PBYs in the East Indies were called *Y-Boats* from their serial prefixes (cf *X-Boat*).

Canada dropped its original name of Convoy for the PBY to avoid the possibility of confusion with shipping convoys, and in 1941 the RCAF opted for the names Canso and Canso A (after the Canso Strait, Nova Scotia) for flying boat and amphibian models respectively; contrary to popular belief, Canso is not a blanket term for all amphibious models, nor does it apply to Canadian-built aircraft in foreign service. The PBN-1 model built by the Naval Aircraft Factory at Philadelphia with a taller fin and improved hydrodynamics was renamed **Nomad**, and the sole PBY-5R Sea Mare was a transport with extra cabin windows but no nose turret.

The long range of the PBY was achieved at the expense of low power, and a post-war attempt to remedy this

resulted in the Steward Davis Super Cat conversion with 1,700hp Wright Cyclones removed from surplus NA B-25 Mitchell bombers. A more ambitious Steward Davis project was the unbuilt Skybarge of 1965 with optional extra power provided by a Jet Pak above the wing (cf *Dollar 19*) and fore-and-aft cargo bays, each with their own 4,000lb hoist. Southern California Aircraft's Landseaire PBY-5A conversion was an ultra-luxurious air-yacht with two-crew operation, sleeping accommodation for eight people, a pair of 14ft sailing dinghies slung under the wings, and even retractable airstairs designed to double as a diving-board when the aircraft was afloat. Perhaps the most radical of all PBY conversions though was the 1968 Bird Innovator (N5907) fitted with modern 340hp Lycoming inlines outboard of the existing P&W Twin Wasp radials, making it the world's only four-engined Catalina. The sole example served the Bird Corporation itself (a producer of medical equipment) for many years, and was restored to flying condition as N5PY in 1991.

Catfish

Occasional nickname for the Sikorsky UH-60 **Blackhawk** transport helicopter and its descendants, after their flat-bottomed and slightly humpbacked look. The UH-60 (also known to its makers as the S-70) is the second Sikorsky helicopter to bear the **Blackhawk** name, following the private-venture S-67 tandem-seat gunship, the sole example of which crashed during the 1974 Farnborough air display. Designed in the early seventies to replace the ubiquitous *Huey* and given such an adaptable name by the US Army, it was perhaps inevitable that the **Blackhawk** should spawn a whole raft of individually named variants. Apart from the UH-60 itself, mainstream variants include the US Navy's SH-60B Seahawk (S-70B) and the less sophisticated SH-60F Ocean Hawk 'CV-Helo', and the US Coast Guard's HH-60J Jayhawk. The USAF's HH-60D Night Hawk combat rescue version was cancelled on cost grounds after a single prototype flew in 1984, and as an interim replacement basic 'vanilla' UH-60s were modified with provision for inflight refuelling as Credible Hawks; these in turn were later updated as MH-60G Pave Hawks with a similar Pave Low III navigation system to that of the Sikorsky MH-53J.

Further variants include the US Army's EH-60C (formerly EH-60A) Quick Fix II electronic counter-measures version, the specially configured S-70A Desert Hawk fifteen-seat utility model for the Royal Saudi Land Forces, and the 24 S-70C-II Hei Ying (Sea Eagle) transports with under-nose radar delivered to China in 1984/85. Taiwan's fourteen S-70C rescue helicopters are nicknamed *Blue Hawks* after their striking blue and white colour scheme, and four US Army **Blackhawks** loaned to the Customs Service to interdict drug-running aircraft around Florida were given the appropriate but strictly unofficial title of *Pot Hawks*.

Centurion
The stylish Miles M100 Student is often remembered as one of the best trainers the RAF never had, since by the time it flew in May 1957 the Service was already committed to the Jet Provost. Projected developments included the slightly enlarged Centurion trainer and the Graduate communications aircraft, both of which were four-seaters, and as late as 1983 the Merlin was being offered as a Jet Provost replacement (to AST 412) with new tandem seating, repositioned intakes, and a single Garrett F109 engine.

Century Series
Well-known collective term for the new wave of all-supersonic USAF fighters, beginning with the F-100 Super Sabre and ending (in nomenclature terms at least) with the 1962 joint designation system. Although it could conceivably be stretched to take in the General Dynamics F-111 and even the Lockheed F-117, the term is generally accepted as applying to just six types, namely the F-100 *Hun*, F-101 *One-O-Wonder*, F-102 *Deuce*, F-104 *Bullet & Blades*, F-105 *Thud*. and F-106 *Six Shooter*. Of the 'missing' designations, the F-107 *Super Super Sabre* was cancelled, the Republic F-103 **Thunderwarrior**, North American F-108 Rapier and VTOL Bell F-109 did not reach the flight test stage, and the F-110 Spectre became the F-4C Phantom. It is often claimed that the Hughes Falcon and Boeing Bomarc missiles were given their F-98 and F-99 designations purely to bring forward the landmark F-100 title for use on the world's first genuine supersonic fighter. In fact the Bomarc was designated F-99 in 1949 whilst the Super Sabre did not become the F-100 until 7 December 1951. This would seem to suggest that rather than being brought forward, the F-100 designation was kept on ice for the first supersonic type, as it seems unlikely that there were no other USAF fighter projects requiring a designation in the intervening period.

Chain Fighter

Spanish Nationalist forces in the civil war frequently referred to their Heinkel He 51 biplanes as the Cadena (from Caza de Cadena, or *Chain Fighter*) after their successful tactic of line-astern, ground-strafing swoops.

Cherokee

Piper began adding model names to its famous PA-28 **Cherokee** line with the introduction of the Cherokee Arrow in 1967 (unless one counts the stretched PA-32 Cherokee Six [not '6'] of 1965) and from 1973 onwards all variants in production had double-barrelled names. However, from 1978 the company completely dropped this most famous of all lightplane names (in the absence of basic type names for its Cessna rivals) leaving the then-current model names to stand alone; thus the Cherokee Arrow III (for example) became simply the Piper Arrow III and later developments such as the Dakota, Saratoga, and Cadet have no formal connection with the **Cherokee** title. Several PA-28/PA-32 models were also licence-built in Brazil by Embraer and their Neiva associate, with the following names: Carioquinha (Archer II), Corisco (Arrow II), Carioca (Pathfinder), Tupi (a stripped-down Carioca/Pathfinder), Minuano (Six) and Sertanejo (Lance).

Chetak

Indian Air Force/Navy name for the Aerospatiale SA 316 Alouette III helicopter, built under licence by HAL (Hindustan Aeronautics Ltd) at Bangalore from 1965; the name comes from the horse of Maharana Pratap, the warrior king of Rajputana, and should not be confused with the Cheetah name applied in India to the **Lama** helicopter. Among other

military users of the best-selling Alouette III was Rhodesia, where the air force had a local adaptation called the *K-Car* with a 20mm machine-gun mounted on a strengthened cabin floor and firing through a door aperture; unarmed troop-carrying Alouette IIIs were known as *G-Cars* and the names may have signified 'Killer' and 'Grunt' (soldier) carrying versions. In South Africa the French machine is popularly known by the diminutive *Alo*, and in February 1985 Atlas Aircraft first flew the extensively redesigned XH-1 Alpha tandem-seat gunship derivative; no further conversions were carried out, but the Alpha (along with experimental Puma versions) was useful in validating gunship technology later applied to the larger CSH-2 Rooivalk (Rock Kestrel) helicopter. A tandem-seat light-attack development of the locally-built Alouette III also appeared in Romania in 1984 as the ICA IAR-317 Airfox; after an early cancellation the project was revived in 1990, and about twelve currently serve with the Romanian Air Force.

China Clipper

China Clipper was actually just the name of the first of Pan Am's trio of Martin M-130 flying boats (the others being Philippine Clipper and Hawaii Clipper) but with the ballyhoo of the inaugural San Francisco-Manila flight (22 November 1935) and a Humphrey Bogart film of the same name, the title came to be regarded as a kind of unofficial class name for all three aircraft.

Chippie

Also known as the *Flying Sardine* because of its slim fuselage, the de Havilland Canada DHC-1 Chipmunk primary trainer first flew at Downsview, Ontario in May 1946, but the lion's share of production (exactly 1,000 aircraft) was undertaken by the parent company in the UK. Large numbers became surplus in the fifties (particularly RAF T10s) and in time formed the basis of a number of widely varied conversions. The so-called 'Masefield Variant' (after Peter Masefield) featured the Canadian-style blown hood plus wheel spats and luggage bays in the wings, whilst in Australia there appeared the Sasin SA-29 Spraymaster, which was a single-seat agricultural aircraft broadly similar to DH's own Chipmunk 23; a further single-seat Australian conversion, circa 1966, was the Aerostructures Pty Sundowner, a sports model with a new 180hp Lycoming motor, blown hood and tip-tanks.

One-off conversions of Canadian airframes have included the Cheekee Chipmunk CF-CYT with a 210hp Continental and wheel spats, and Art Scholl's well-known Super Chipmunk (N13Y) single-seat competition aerobatics aircraft, with redesigned tail, clipped wings, retractable gear and 280hp Lycoming; the latter apparently also provided the inspiration for Nigel Brendish's Mighty Munk (G-IDDY), which in 1981 briefly made the news by flying a 72nd anniversary reconstruction of Blériot's Channel crossing – inverted. Fifteen Royal Thai Air Force Chipmunks were modified in the early eighties with a new 180hp Lycoming, enlarged 'Sphinx' tail and redesigned hood, to become the RTAF-4 Chantra (Moon).

Chirri
Almost 400 of the supremely agile Fiat CR32 fighter biplanes served with Italian and Spanish Nationalist forces in the Spanish Civil War, where they received the local nickname *Chirri*. The name was also used, more or less officially, for the 100 CR.32quater (Mk 4) machines built under licence by Hispano as the HS-132-L.

Chitral
Original choice of name for the Handley Page Clive, a bomber-transport derivative of the twenties-vintage Hyderabad/Hinaidi bombers. Just three Clives were built, two with metal airframes going to the RAF in India, and the original wooden Clive 1 to Sir Alan Cobham.

Chopper
Chopper soon replaced early American names like *Eggbeater* and *Frustrated Palm Tree* as a generic term for helicopters, but whilst it may be prevalent in both the US Army (the world's biggest operator of helicopters) and USAF, *Helo* is used almost exclusively in the US Navy and Marine Corps. RAF counterparts, though less common, are *Hydraulic Palm Tree* and the echoic *Wokka*, whilst in the Army Air Corps (AAC) transport helicopters are *Cabs*.

Chuff
Grumman's new TBF torpedo bomber was allegedly given its emotive title of Avenger on the day the Japanese attacked Pearl Harbor (news of which also brought an abrupt end to the Company's annual Open House), but as it went on to become the backbone of US Navy strike wings its portly lines (so typical of the Grumman 'Ironworks') earned it such names as *Chuff, Turkey* and even *Pregnant Beast* after the Curtiss SB2C *Beast*. Inevitably, the TBM-3W Airborne Early Warning model with its bulbous

ventral radome became yet another
Guppy. No less than 7,546 out of a
grand total of 9,836 Avengers were
TBMs built by General Motors'
Eastern Aircraft Division, who at one
stage had ambitions to name their
aircraft Sea Eagle, whilst the Royal
Navy knew the type as the Tarpon
until early 1944.

Cityliner
Name adopted for the Saab (formerly
Saab-Fairchild) SF340 twin turboprop
by the Swiss airline Crossair, who in
June 1984 became the first to put the
35-seater into service.

Clansman
Intended class name for BEA's
planned fleet of Handley Page (ex-
Miles) Marathon feederliners,
ordered as Dragon Rapide
replacements but cancelled in 1952
after just a single trials aircraft (G-
ALUB Rob Roy) had appeared in the
Corporation's colours.

Clay Pigeon
Name originally proposed for the
attractive Airspeed AS 30 Queen
Wasp radio-controlled target drone
biplane. The Queen Wasp was
criticised in its day for being too
expensive for the role and just seven
were built from 1937, in both
landplane and floatplane forms.
Proposed variants were the AS 38
communications aircraft and the AS
50 trainer, which would presumably
have dropped the **Queen** prefix to
become simply the Airspeed Wasp.

Cleveland
The Hollywood-inspired name of
Helldiver was clearly not to British
tastes: despite receiving just five
Curtiss SBC-4 biplanes and quickly
relegating them for use as RAF

instructional airframes, the British
chose to rename them **Clevelands**.
Though widely used by Curtiss for
both the SBC and the earlier F8C/02C
series, the Helldiver name was not
formally recognised by the US Navy
either, until the monoplane SB2C
Beast.

Clipper
Former name of Cessna's all-new
Model 303 light twin, first flown in
1979, but changed to Crusader in
deference to Pan Am's registration of
the name.

Cloudmaster
Name used only very briefly for the
Douglas DC-6 propliner. Although
Liftmaster is often quoted as the name
of the military C-118 and R6D, it was
actually coined for the stretched DC-
6A freighter on which they were
based. Pan Am acquired a fleet of 45
DC-6Bs specially configured for high-
density tourist operations, which they
called the Super 6 or Super 6 Clipper.

Clunk
Destined to be Canada's only real
indigenous warplane to go into
production and service, the twin-jet
Avro Canada CF-100 Canuck all-
weather fighter was fondly known to
its crews as the ***Clunk***, ***Lead Sled***,
Beast or (from the zeroes in its
designation) the *Zilch*. The name
Jaeger (Hunter) was associated with
the Mk 4 around 1956, possibly as a
result of interest from West Germany.
A total of 692 copies of the underrated
CF-100 were eventually built, and the
type also served with the Belgian Air
Force.

Flying Coathanger
Descriptive term for the Rutan
VariEze homebuilt aircraft with its

long swept wings extending gracefully
behind a short, tailless fuselage.

Cobra
Northrop named its P-530 fighter
design after the 'Cobra Hood' Leading
Edge Root Extensions (LERXs) to its
wings, but the title was set aside for
the pair of derivative YF-17s that lost
the USAF's Lightweight Fighter
(LWF) fly-off to the General
Dynamics F-16 in 1975. From the
broad base of the YF-17 however,
McDonnell Douglas and Northrop
evolved the F/A-18 Hornet for the US
Navy and there were also plans for a
de-navalised F-18L for export, known
tentatively by Northrop as the Cobra
II; with its greater experience of naval
aviation McDD had 60 per cent of the
F/A-18 programme to Northrop's 40
per cent with the percentages reversed
for the land-based model. The F-18L
would have saved at least 2,200lb of
weight by lightening the undercarriage
and deleting the wing-fold mechanism,
but in all export competitions it was
edged out by the off-the-shelf F/A-18
Hornet from partners McDonnell
Douglas, resulting in the famous
lawsuit between the two companies;
the action was settled out of court in
April 1985 with a payment to Northrop
and an agreement that McDD should
be prime contractor on *all* F/A-18
derivatives. The F/A-18 itself was at
one stage nicknamed the *Star Wars
Fighter* after its advanced
computerised cockpit displays,
although the name proved short-lived.
The first foreign customer for the F/A-
18 was Canada, where it is designated
CF-188 and is officially nameless,
Hornet being deemed unsuitable
because of its non-bilingual nature.

Coffee Pot
Like the better known Breguet 14

bomber of WW1 (cf **Limousine**) the
original 1909 Bre 1 biplane was also
known as the *Coffee Pot*; the origins
are less clear, but its 'rectilinear' form
has been suggested.

Flying Coffin (i)
As the Luftwaffe's first real bombers,
the lumbering Dornier Do 11, 13 and
23 series featured in many an upbeat
Nazi propaganda film, but to their
crews they had an abysmal reputation.
Under certain conditions the wings
seemed determined to shake
themselves to pieces, stability and
structural integrity both left much to
be desired, and the crude retractable
undercarriage of the Do 11 was so
unreliable it had to be permanently
locked in the down position.

Flying Coffin (ii)
Swedish Air Force epithet for the
unreliable Caproni Ca 313, an Italian
light-bomber/recce aircraft that was
actually ordered for the RAF in 1940.

Colibri
Colibri (Humming Bird) was the
original and apparently short-lived
name of the current Franco-Italian
ATR 42 twin-turboprop regional
airliner. Maritime patrol variants of
both the ATR 42 and stretched ATR
72 were announced in 1988 as the
Petrel 42 or 72, armed with torpedoes
or Exocet missiles, and equipped with
much of the avionics suite of the
Dassault Atlantique 2.

Commodore Jet
Israel Aircraft Industries (IAI)
ultimately have the American anti-
trust laws to thank for their successful
corporate jet line, for when North
American merged with Rockwell in
1967 the huge new combine was
disbarred from producing two

34

competing types; the N.A. Sabreliner was the better established of the pair (particularly as the military T-39) and so the Rockwell Jet Commander 1121 programme was sold off. IAI began by selling the remaining 49 American-built aircraft (out of a total of 149) as Commodore Jet 1121s, and the first true Israeli model was the Commodore Jet 1123 with a 22in fuselage stretch, uprated engines, tip-tanks, and various wing modifications; deliveries began in 1972 as the Commodre Jet Eleven-23, but barely a year later the name was changed again to 1123 Westwind, to emphasise its difference from the American models.

The Israeli Navy's trio of 1124N Sea Scans are armed and comprehensively equipped coastal patrol aircraft with Litton APS-504V radar in a bulbous nose radome, and are based on the further improved 1124 Westwind, which introduced Garrett TFE-731 turbofans in place of the earlier GE CJ610 turbojets. The Australian Customs Service also has three Sea Scans (known as the *Platypus*) equipped with FLIR (Forward-Looking Infra Red) and operated under the Coastwatch law enforcement programme, hence the occasionally used title Coastwatcher. The IAI 1125 Astra (briefly the Westwind 1125) of 1984 was a radical departure with swept wings, and the forthcoming Galaxy will bear even less relationship to the Commodore/Westwind line.

Concord
Contrary to some accounts, BAC and Sud Aviation readily agreed that the name (first suggested by the son of BAC's publicity manager) of their prestigious supersonic transport should be spelt with a Gallic 'e', only

for the Conservative Government of the day, piqued at not being consulted and angered by de Gaulle's 'Non' to Britain joining the Common Market, to insist on the English spelling. BAC's George Edwards characteristically declined to go back on his word, and so for five years both versions were in use in the UK. However, by the time of the infamous sub-zero roll-out of aircraft 001 at Toulouse on 11 December 1967 there had been a General Election in Britain and it was there that Anthony Wedgewood Benn formally recognised the contentious 'e' on behalf of HM Government. International collaboration has since come a long way.

Concordski
Tabloid press name for the Soviet Tupolev Tu-144 (NATO – Charger) Supersonic Transport (SST), particularly around the time of the crash at the Paris Air Show on 23 June 1973.

Condor
Traditional name for Curtiss transport aircraft, except for the twin-engined CW-20 Condor III monoplane dropped in favour of Commando when ordered in quantity for the USAAF as the C-46. Occasionally referred to as the ***Whale*** on account of its size and bulk, over 3,000 C-46s and Navy R5Cs were built, but attempts to adapt it to post-war civil operations were plagued with certification problems, and aircraft modified to the CAA's T-Category specification became known informally as ***T-Cats***.

Connie
With the possible exception of Concorde, no other airliner has had such widespread aesthetic appeal as

the Lockheed Constellation, a true classic that epitomised the golden age of propliners in the fifties. Known in the project stage as the Excalibur A and initially put into production as the USAAF's C-69 (later military versions included the C-121 and R7V) the first purely civil post-war derivative was the 1946 L-649 model promoted by Eastern Air Lines as the Gold Plate Constellation to emphasise its greatly improved cabin fittings; Gold Plate Special was Eastern's term for the L-749 with increased fuel. An all-freight model of the projected L-949 would have become the Speedfreighter, but instead effort was switched to the L-1049 Super Constellation with an 18ft 5in fuselage stretch and (from the 25th aircraft) Wright Turbo-Compound engines giving a 40 per cent increase in payload; principal civil versions were the L-1049C, L-1049G (the *Super G*) and the convertible L1049H *Husky*. The ultimate commercial variant, and indeed the ultimate piston airliner, was the L-1649 Starliner or *Super Super Constellation*, with an entirely new constant-taper wing and further 2ft 7in stretch over the L-1049 series; overtaken by jets, only 44 were built, becoming known as the Super Star by Air France and the Jetstream class by TWA.

The US Navy were the first to test an Airborne Early Warning variant with their pair of L-749A-based PO-1Ws (*Po One*) but all subsequent AEW machines were L-1049 Super Connie derivatives. Although the title Warning Star officially applied to both the US Navy and Air Force AEW models, the former's WV-2 was equally well-known by the colloquial *Radome*. *Elation* was another Lockheed hybrid nickname for a US Navy R7V transport retained to prove

the Allison 501 turboprop installation for the L-188 Electra airliner.

Flying Cooling Tower
Nickname for the experimental SR-N1 (Saunders Roe, Nautical No. 1) hovercraft, after the prominent vertical intake duct to its Alvis Leonides engine. The name is believed to have originated around the time of the 1959 Farnborough show when, wearing the Class B registration G-12-4, it demonstrated the potential of the hovercraft by carrying twenty armed troops on its open deck.

Flying Corkscrew
Name coined by test pilot Louis Paget for the Westland F20/27 Interceptor (J9124) monoplane of 1928, describing its odd tendency to roll itself level at the top of a loop, and its sluggishness in coming out of a spin.

Corporate 77
Collective title used by Boeing between 1985 and 1987 for executive/VIP versions of its jet airliner family. The name was used in conjunction with the standard model number; thus a VIP 737-300 was a Corporate 77-33, and a VIP 757-200 a Corporate 77-52.

Corsair
Cessna's classy Model 425 turboprop twin was known as the Corsair for the first three years of its production life, but was renamed Conquest I in 1982; at the same time the larger Model 441 Conquest became the Conquest II.

Cow
Triton was the formal name applied by the Irish Army Air Corps to the trio of A.S. Mongoose-powered Avro 621 trainers received in 1930, but to its crews the underpowered biplane was invariably known as the *Cow*. The

RAF also had a trial batch of
Mongoose-engined 621s, but it was
with a more powerful A.S. Lynx that
it became famous as the Tutor, with
almost 400 built for the Service up to
1936. The unnumbered Avro Prefect
was a minor variant for navigation
training, with just seven going to the
RAF and four to New Zealand.

Cradle of the Air Force
Contrived nickname applied to the
USAAF's Fairchild PT-19 Cornell
primary training monoplane of WW2.

Cranberry
Occasional RAF nickname for the
famous English Electric/BAC
Canberra twin-jet bomber. The policy
of naming bombers and transports
after British Empire towns began to
fall into disuse after WW2 (as
exemplified by the Sperrin,
Washington, and the **V-Bombers**) and
it was speculated at the time that the
Canberra name was chosen to help
kindle RAAF interest, and/or to
honour the Australian contribution to
Bomber Command in the war.
    Among the many achievements of
the Canberra was the remarkable sale
to the USAF and licence-production
by Glenn L. Martin as the B-57. The
agreement stipulated that Martin
should also call the B-57 the Canberra
and employ its 'best efforts' to
persuade the US Government to do
likewise; although both Martin and
the USAF formally recognised the
name it gained little real currency in
the US, and publicity material
describing the role of the B-57 gave
rise to *Intruder* and *Night Intruder*,
both of which are sometimes mistaken
for official USAF titles. In south-east
Asia the Vietcong knew the B-57 by
the derogatory term *Caterpillar*, and
Nighthawk was the unofficial name of

sixteen B-57Gs with radar, Low Light
Level TV (LLLTV), Infra-Red, and
other sensors.

Crane
With a few notable exceptions, Soviet
warplanes were not usually given
names, but since the collapse of
Communism a number of semi-official
titles have begun to appear in
connection with Russian Air Force
aircraft, notably Sukhoi types. Among
them is Zhuravlik or *Crane* for the
well-known Su-27, although improved
relations and increased
communication with the West has
meant that the Su-27 is also now
widely known in its homeland by its
NATO codename of Flanker. The Su-
27 is of course broadly comparable
with the McDonnell Douglas F-15
Eagle, and like its American
counterpart also exists in two-seat
strike form; intended to replace the
Su-24 (NATO – Fencer), the Su-27IB
(alias Su-34) has side-by-side seating
giving it a new broad nose shape, and
the almost inevitable nickname of
Platypus.

Flying Crash
The outrigged helicopter, in which the
rotors are mounted at the ends of
wings or booms, surely reached its
zenith with the massive Russian Mil
Mi-12 (NATO – Homer) first revealed
to the West at the 1971 Paris Salon.
Span over the twin five-blade rotors
was no less than 219ft, and with a mass
of struts and undercarriage members
supporting the inversely-tapered
wings, the Mi-12 certainly deserved its
unusual nickname.

Crikey
Rather like the American 'Stealth
Fighter' four decades later, the
Westland Whirlwind twin-engined,

single-seat fighter was one of the biggest open secrets of its day. Schoolboys built models of a type that until February 1942 didn't officially exist (unlike those early 'Stealth' kits they were usually accurate), the Germans had it in their recognition books, and people living around Yeovil at the time even had a nickname for it; *Crikey*, borrowed from the Shell petrol advert in which a farm worker, startled by a speeding car, exclaimed 'Crikey, that's Shell, that was!' The *Whirligig*, as it was also known, was used by only two RAF squadrons (137 and 263) largely as a low-level fighter-bomber; hence *Whirlibomber*.

Criquet
Having built the Fieseler Fi 156 Storch (Stork) STOL communications aircraft under German occupation since 1942, the Morane-Saulnier works at Puteaux (Seine) continued production post-war as the MS500 Criquet (Locust). A similar situation arose at the Mraz works in Czechoslovakia, where the post-war model was designated K-65 Cap (Stork).

Crocodile
Appropriate nickname within the Soviet/CIS air forces for the MiG-23 (NATO – Flogger) swing-wing fighter, with its angular, predatory appearance; the flattened laser-rangefinder nose of the closely related MiG-27 ground-attack aircraft (also featured on the MiG-23B and BN) has earned it the widespread nickname of *Utkanos* or *Duck Nose*. In the Indian Air Force the Mig-23BN (Flogger H) is known as the Vijay (Victor), the MiG-23MF (Flogger B) as the Rakshak (Guardian), and the MiG-27M (Flogger J) built locally by

HAL has the name of Bahadur (Valiant).

Cross-Eyed Monster
Known as the HR2S *Deuce* to the US Navy and H-37 Mojave to the Army, the piston-engined Sikorsky S-56 heavylift helicopter featured two huge shoulder-mounted engine nacelles with a curious canted-in appearance when viewed from head-on. A temperamental and noisy beast, the fearsome visage was completed by a 'mouth' of horizontal windows in the blunt nose.

Cuauhtemoc
Title of the Maule M-4 lightplane variant built by Servicios Aereas de America SA of Mexico, with a new 180hp Lycoming in place of the standard 145hp Continental. The first all-Mexican version of the 'Mighty Maule' flew in April 1964, but the project seems to have foundered after only about three machines.

Curious Ada
Play on the name of the Short-Bristow Crusader racing seaplane. Also known as the *Blind Wonder* because of its poor forward visibility, the twin-float monoplane crashed on take-off (due to crossed-over aileron controls) whilst being used as a practice machine before the 1927 Schneider contest at Venice.

Dankok
Name given to the dozen Hawker Woodcock II fighter biplanes built by the Danish Royal Naval Dockyard in 1927/28, following the acquisition of three British-built examples known as Danecocks.

Dark Shark
Intriguing but unofficial name given to

38

the one-off Ryan XF2R-1 fighter of
1946, an experimental version of the
US Navy's composite power (jet and
piston engines) FR-1 Fireball with a
General Electric XT31 turboprop in
place of the usual radial engine.

DC-Twee
Nickname of five Northeast Airlines
Douglas DC-2 airliners fitted with
DC-3 wings during WW2.

Debonair
The trademark 'V' (or 'Butterfly') tail
of the classy Beech Bonanza tourer
(first flown way back in 1945) was not
to everyone's taste, so from 1959
Beech offered a version with a
conventional three-surface tail which
was known as the **Debonair** until 1967
when it was redesignated Model 33
Bonanza. A conversion package
offering a conventional tail was also
available from Mike Smith Aero in the
mid-eighties as the Tri-Tail Bonanza.
Colemill Enterprises retain the V tail
on their Starfire conversion however,
which boasts a 300hp Continental IO-
550B engine, four-blade Hartzell
Sabre Blade prop and 'Zip-Tip'
winglets amongst its improvements.
    The most radical of all Bonanza
conversions though must be the twin-
engined Super V ('vee', not 'five')
which also kept the V tail but had the
original single motor replaced by a
new fibreglass nose and a 170hp
Lycoming mounted on the leading
edge of each wing in a remarkably
simple installation. Originating with
the Peterson Skyline Co. of Oklahoma
in 1955, actual conversions (believed
to total fourteen) were carried out by
Oakland Airmotive and Fleet Aircraft
of Ontario.

Deffy
Deffy and *Daffy* were both nicknames

for the RAF's Boulton Paul Defiant,
the single-engined turret fighter that
was withdrawn from daylight ops in
August 1940 and successfully adapted
for night-fighter duties.

Delfin
Provisional name (meaning 'dolphin')
for the Argentinian FMA IA 58
Pucara (Fortress) twin-turboprop,
counter-insurgency aircraft.

Delta Mystère
Original and somewhat misleading
title of the tiny Dassault MD 550
Mirage I of 1955. Powered by a pair
of licence-built Armstrong-Siddeley
Viper 5 turbojets, the type was
originally conceived as a technology
demonstrator for the larger twin-
engined Mirage II project (unbuilt)
and had little in common with the
ultimate Mirage III series.

Delta Queen
Occasional nickname for the Convair
B-58 Hustler, the USAF's first
supersonic bomber. Characterised by
its delta wing, four huge podded
engines, and external mission pod, the
B-58 was withdrawn in 1970 after a
relatively short service life of just ten
years.

Deuce
Originally they thought of calling it
Machete, officially it became the
Delta Dagger, but to its crews the
Convair F-102 interceptor was never
anything but the *Deuce* (a two at cards
in the US). The original YF-102 was
of course famous for being salvaged by
the NACA's Area Rule aerodynamic
formula, but the early nickname of
Hot Rod for the recontoured YF-
102A model was quickly
overshadowed by more graphic
allusions to Coke bottles, Marilyn

Monroe and 'Love Handles'. Nothing could give the TF-102A trainer a genuine supersonic capability though, and with its draggy side-by-side cockpit the *Delta Dud* or *Pig* was one of the least representative of all conversion trainers. In the seventies over 200 obsolete F-102s were converted into drones under the Sperry Pave Deuce programme, hence their final nickname of *Spad*.

Devil's Broomstick
Graphic term coined by test pilot Konstantin Gruzdev for the Soviet Union's tiny Bereznyak-Isaev BI rocket-propelled interceptor, first flown in May 1942. A top speed of no less than 600mph at sea level was projected for the BI, but production was abandoned after a fatal crash and the discovery of incurable aerodynamic problems.

Devil's Chariot
Title used by the Mujahedin fighters in Afghanistan for the hated Soviet Mil Mi-24 (NATO – Hind) assault support/gunship helicopter, known more fondly to its crews as the *Gorbach* or *Hunchback*.

Devon
RAF name for the de Havilland DH 104 Dove twin-engined light transport, along with Sea Devon for the Royal Navy version. The Dove feederliner first flew back in September 1945 and though it sold to the tune of 542 copies until 1967, DH and Hawker Siddeley have often been criticised for not exploiting its development potential to the same extent as Beech, for example, did with their **Queen Air**/King Air line. In 1958, however, it was reported that DH were studying a stretched development with a swept fin, weather

radar, and 520hp Lycomings in place of the usual DH Gipsy Queens; the project was said to be named Diplomat, perhaps to follow on from the titles of the former Airspeed company, which DH had finally absorbed seven years previously. Unfortunately the slightly mysterious Diplomat never became hardware, but a broadly similar conversion package emerged in the US in the sixties as the Riley Executive 400 with a standard-length fuselage, 400hp turbo-supercharged Lycomings, and optional swept fin; at least twenty conversions were carried out, including four by McAlpines in the UK, but a projected Turbo-Executive 400 with Turbomeca Astazou turboprops failed to materialise. As such, the only known turbine Dove was the Carstedt CJ600A Jet Liner, with Garrett TPE-331s and an 87in fuselage stretch providing seating for up to eighteen people; the first of six conversions flew at Long Beach, California in December 1966 (cf *Double Dove*).

Diana
Quite possibly the world's most beautiful biplane (at least in the definitive dual-cockpit version), the de Havilland DH 86 four-engined airliner operated with Imperial Airways (and its associate Railway Air Services) under the class name of **Diana**, although the flagship itself, G-ACPL 'Diana', was quickly rebuilt with the dual cockpit and renamed 'Delphinus'; with Qantas Empire Airways the DH 86 was known as the Commonwealth class. In the project phase the type was occasionally referred to informally as the Dragon Express (denoting its relationship to the DH 84 Dragon twin) and until at least 1937 de Havilland maintenance

documents referred to it as the Express Air Liner, but despite claims to the contrary it seems unlikely that either was ever really intended to be a proper name. The end-plate fins introduced with the DH 86B were colloquially known as 'Zulu Shields'.

Diesel 8

Most jet airliners not only tend to look alike, but few even have a name to help distinguish one from another. Douglas called their all-freight or combi DC-8F the Jet Trader, but otherwise left it to the airlines to address the anonymity question if they so wished. Eastern Air Lines were quick off the mark with the name Golden Falcon (and the unofficial designation DC-8B) emblazoned across their DC-8-21s, the Flying Tiger Line pre-empted the 747 by briefly labelling its Super 63s Jumbo Jets, and Canadian Pacific knew its 199-seat Super 63s as Spacemasters.

Digby

Royal Canadian Air Force name for the Douglas B-18 Bolo, a rather lacklustre twin-engined bomber based loosely on the DC-2 airliner.

Dog

The North American F-86D all-weather Sabre was widely known as the *Sabredog*, *Dog* or *Dogship*, but it was much more than simple phonetics, as the new nose radome and chin intake gave it an unmistakably canine visage. However, the suffix letter of the NATO F-86K model led to one of the quaintest of all aircraft nicknames, and one that could only have come from the Netherlands: *Cheesefighter*, after the former Amsterdam Superintendent of Police Mr Kaasjager (literally 'cheesefighter' or - 'hunter'). In 1964 the Japanese Air

Self-Defence Force (JASDF) adopted the local names Kyokko (Aurora) for the F-86F and Gekko (Moonlight) for the F-86D, but neither seem to have gained much popular usage. From the F-86H onwards the Sabre name was formally styled Sabre Jet by North American and the USAF.

Sea Sabre may seem an obvious title for the naval F-86 variants but, allegedly for political 'Funding Procedures' reasons, the US Navy resurrected the name of the old straight-winged FJ-1 Fury, from which in fact the whole Sabre line originally evolved. The name Sea Sabre, incidentally, was used for a 1975 T-39 Sabreliner proposal to the US Coast Guard. Whilst the FJ-2 and FJ-3 Furies were clearly recognisable as hooked Sabres, the FJ-4 amounted to virtually a new aircraft, but all were known within the US Navy as *Tinker Toys*, *Scooters* or *Mighty Mites*.

Dogan

Bulgarian Air Force name for the compact Avia B534 Czech fighter biplane, at least 48 captured examples of which were passed on by Germany.

Dog Ship

Old American expression for an aircraft retained by the manufacturers for development work, for example the famous 'Dash-80' prototype used by Boeing for eighteen years in support of its various jet airliner programmes.

Dollar 19

Strickly speaking, the Fairchild C-82 military freighter was the Packet, and the C-119 development with a relocated cockpit was the **Flying Boxcar**, but confusion was perhaps inevitable from the moment the prototype XC-82 Packet was rolled out in 1944 with the legend *Flying Boxcar* on its nose. Both also shared the nickname of *Crowd Killer*, from the number of troops they could accommodate (42 paratroops was standard in the C-82). The twin-boom layout and long production run (over 1,000) of the C-119 made it ripe for development and conversions, and one-offs included the 1950 XC-120 Pack-Plane with a detachable under-fuselage cargo pod, and the 1952 YC-119H Skyvan which featured 40 per cent larger wings and redesigned tail and booms. In the twilight of the *Dollar 19*'s career, 26 C-119Gs were hurriedly converted into AC-119G Shadow gunships with a quartet of 7.62mm GAU-2 miniguns firing from the port side for use in Vietnam, and were followed by a similar number of the definitive AC-119K Stinger version with an additional pair of 20mm M61A1 guns and underwing GE J85 booster jets.

Jet augmentation of the C-82 and C-119 with 'Jet Paks' (typically a dorsal-mounted Westinghouse J34) was also one of the specialities of Steward Davis Inc. of Long Beach, who marketed conversions as the Jet Packet and Jet Pak C-119; among aircraft that were modified was the Indian Air Force fleet of C-119Gs. From their Jet Packet II, Steward Davis went on to develop the reduced weight Skytruck, and although a C-82 appeared as a Skytruck in the 1965 film *The Flight of the Phoenix*, the original novel by Elleston Trevor carefully referred to it as 'Salmon-Rees' Skytruck. The most extensive Steward Davis development of all was the Skypallet, in which the cargo-hold floor doubled as a detachable and fully roadworthy truckbed, raised and lowered from the hold by a 40,000lb hoist; a motorised and steerable

nosewheel enabled the crew to position the aircraft over a load and pick it up without ground assistance and without having to leave the cockpit.

Dolphin
US Coast Guard name for its HH-65A rescue helicopter, a customised variant of the Aerospatiale/Eurocopter AS 365N Dauphin 2; designated AS 366G by the manufacturers, it features approximately 60 per cent (by value) American equipment, and a total of 99 were delivered between 1984 and 1989. Harbin Aircraft Manufacturing Co. (HAMC) of China also assemble the AS 365N as the Z-9 Haitun (Dolphin). The AS 565 Panther is a dedicated multi-role military variant of the Dauphin 2, first flown (as the AS 365M) in 1984.

Double Dove
Very early nickname for the de Havilland DH 114 Heron, the stretched, four-engined development of the DH 104 Dove twin (cf **Devon**). At different times BEA operated three fixed-gear Heron 1Bs in Scotland as their Hebrides class from 1955 until 1973. Like its smaller brother, in later life the Heron came in for a number of conversions involving replacement of the original Gipsy Queens with, among others, the Riley Skyliner with 260hp turbo-supercharged Lycoming IO-540s, the Shin Meiwa update of four Toa Airways (Japan) Heron 1s with 260hp Continental IO-470s as the Tawron (from Toa Airways Heron) and, most radically of all, the twin-engined Saunders ST-27 with P&WC PT6A-27 turboprops. Twelve ST-27s were converted in Canada from 1969, but plans for a new-build derivative, the

ST-28, failed to progress beyond a single prototype. Though the name is not strictly applicable to Herons as a whole, the aircraft used by the British Embassy in Washington was known as 'Snow White' because it could just about accommodate seven dwarves . . .

Double Ugly

The mighty McDonnell F-4 Phantom (strictly Phantom II, after the earlier FH-1 Phantom) was many things to many men, but it could never be considered a thing of beauty. In its homeland it was dubbed the *Double Ugly*, *Rhino* or *Old Smokey*, whilst in the Luftwaffe the long nose and wide intakes led to the nickname of *Elephant*. In the early days with the RAF it was the *Tomb* which was a suitably spooky play on the final syllable of the Phantom name, as well as reflecting the notoriously cramped conditions of the rear cockpit. An imaginative McAir employee had suggested the formal name of Mithras after the Persian god of light, and the USAF version was originally to have been the F-110 Spectre, perhaps in an attempt to save face at having ordered a Navy plane; but for the new joint designation system in 1962, it seems likely that most foreign operators would have used the USAF name too. In Israel the F-4 has the alternative name of Kurnass (Hammer).

Downtowner

Name of a projected 40- to 50-seat commercial derivative of the mid-sixties vintage LTV/Hiller/Ryan XC-142 turboprop V/STOL transport, on which the entire wing and engine assembly tilted to provide vertical or horizontal thrust as required. American Airlines and Eastern both showed interest in a civil version for city centre operations, but

performance of the XC-142 was not up to expectations: four of the five prototypes crashed, and the whole Tri-Service programme was cancelled.

Dragon

Name adopted in October 1918 for the planned version of the Sopwith Snipe fighter powered by a 320hp ABC Dragonfly radial engine in place of the Snipe's Bentley BR2 rotary. The much-vaunted Dragonfly proved something of a technical disaster, however, and as a result the **Dragon** never saw service with the RAF.

Dragonfly

Westland cut its teeth in helicopter manufacturing with a licence-built version of the Sikorsky S-51, known to the RAF and Royal Navy (and, by association, many civil operators) as the **Dragonfly**. A further development of the WS-51 was the 1955 Widgeon, with an enlarged five-seat cabin and the rotor-head of the WS-55; fifteen were completed, including three converted WS-51s.

Dragon Lady

Of all the 'Black' programmes from Lockheed's famous 'Skunk Works' (the Advanced Development Projects section), perhaps none captured the public imagination like the temperamental U-2 strategic reconnaissance aircraft, which hit the headlines so dramatically with the Gary Powers shoot-down near Sverdlovsk on May Day 1960. Amongst the more fanciful newspaper terms, *Shady Lady* proved the most enduring, whilst the CIA often referred to it internally as the *Deuce*, from its designation; *Angel* referred of course to the high-altitude performance on which the slow-flying U-2 depended for survival. But to those most closely associated with it, the U-2 has always been the *Dragon Lady*, after the mysterious and fickle oriental from the long-running cartoon strip 'Terry and the Pirates'. Few other aircraft nicknames have ever been so inspired, or so closely matched to the character and role of their subject.

Dumbo (i)
Once-popular American term for any air-sea rescue aircraft, after the Disney cartoon elephant of the same name.

Dumbo (ii)
In contrast to its American rescue counterparts, the experimental Supermarine Type 322 torpedo-bomber of 1943 earned its nickname primarily because its variable-incidence wing was thought to be reminiscent of the Disney elephant's flapping ears, although its portly lines and a tendency towards ever-increasing bounces on landing also added to the overall comic appearance.

Duncan Sisters
Company nickname for the pair of Blackburn CA 15C twin engine transports built in monoplane (G-ABKV) and biplane (G-ABKW) forms for comparative evaluation in 1932. The name honoured both the contemporary Vaudeville stars and designer B.A. Duncan.

Dung Hunter
The deceptively docile nature of the Airco DH 6 trainer and resultant lack of respect (not to mention a number of pranks) was held by many to be the root of its bad reputation in the RFC, reflected in such epithets as *Chummy Hearse*, *Skyhook* and *Crab*.

Paradoxically, though, *Flying Coffin* was first and foremost a reference to the flat-sided, open-topped communal cockpit, and similarly the Australians' favourite term *Dung Hunter* was claimed to have had more to do with the resemblance to a farmyard wastecart than for any tendency to a premature return to earth. Perhaps understandably in view of its reputation, trainee pilots had a habit of panicking and lunging at the over-sized joystick, earning the DH 6 the instructors' title of *The Clutching Hand*.

Flying Dustbin (i)
Derogatory name sometimes applied to the RAF's inter-war Handley Page HP 50 Heyford heavy bomber biplane, after the shape of its retractable ventral gun turret. Instantly recognisable by having its fuselage attached to the underside of the upper wing, the Heyford was sometimes billed as the 'Express Night Bomber', although it did not amount to a proper name.

Flying Dustbin (ii)
Disdainful French term for the indigenous and popular Morane Saulnier/Socata Rallye series of lightplanes, although it must be said that the original 1959-vintage MS 880, in particular, was not a thing of beauty. Waco marketed the MS 894 Rallye Minerva 220 model in the USA as the Sirrius, and in 1977 Pezetel of Poland acquired a licence for the Rallye 100ST which they put into production as the PZL-110 Koliber (Humming Bird), first with a locally-assembled 116hp Franklin and then more successfully with a 150hp Lycoming.

Flying Egg
Nickname of the unusual Siemens-Schuckert DDr I experimental German triplane of WW1, with a short fuselage nacelle (hence the name), fore-and-aft engines, and a tail mounted on outriggers. The sole example crashed on its maiden flight in November 1917, and it was widely decided to abandon plans for a more powerful DDr II version.

Eggbeater
Nickname of the 1946 Curtiss XBTC-2 torpedo bomber prototype with contra-rotating propellers, ie, two sets of blades turning about the same axis but in opposite directions.

Egg Box
Not to be confused with the famous *Tripehound*, the much larger Sopwith LRT Tr three-seat triplane prototype of 1916 had two seats in the fuselage, plus a third in a nacelle on the top wing. Although the nacelle itself was roughly egg-shaped, the name was said to come from the shape of the gunner's enclosure within it.

Elephant
RFC nickname for the large Martinsyde G 100 and G 102 single-seat biplanes, best remembered for their service with 27 Squadron, who adopted an elephant motif in their unit badge. The name is particularly associated with the more powerful G 102 and was apparently restricted to Europe, with the pair being more commonly known elsewhere simply as '*Tinsydes*'.

Empire Boat
Famous but nonetheless unofficial name in the thirties for Imperial Airways' Short S23 flying boats, known to the airline as the C-class (G-

ADHL Canopus was the flagship) and later also as the Imperial Flying Boat.

Ensign Eater
There was something of the look of an unfinished 'Airfix' model about the Chance Vought F7U Cutlass jet fighter, with its truncated rear fuselage, stumpy swept wings, and complete lack of any horizontal tail. Roughly a quarter of the 300-plus built for the US Navy in the fifties were lost in accidents, earning such epithets as *Ensign Eater* (or - *Eliminator*) and the usual *Widow Maker*, whilst the Press looked at the unusually nose-high sit on the ground and came up with *Praying Mantis* and *Pterodactyl*. *Gutless* was more alliterative than descriptive, and despite its poor safety record the radical F7U was not without its devotees.

Eurobus
Swissair name for the Airbus A310 widebody twin. An increasing band of military operators of the A310 includes France, Germany, Thailand and Canada, the latter using the title CC-150 Polaris.

Eurofighter
Long before the controversial European Fighter Aircraft (EFA) or Eurofighter 2000, Saab used this title for its proposed JA-37E, a multi-role Viggen (Thunderbolt) derivative aimed at NATO's F-104 Starfighter replacement contest (the so-called 'Sale of the Century', won in June 1975 by the General Dynamics F-16).

Executaire
Americanised version of the MBB (Messerschmitt Bolkow Blohm, and now Eurocopter) BO 105 light turbine helicopter flown in 1975 by US

distributors Boeing Vertol. Actual conversion work was undertaken by Carson Helicopters Inc., and entailed stretching the fuselage by ten inches, improving cabin furnishings, and installing hinged doors in place of the earlier sliding ones; at least ten conversions were carried out in its first year. A similar stretch appeared in production form as the BO 105CBS, certificated in 1983 and marketed in the US by MBB Helicopter Corp. (who took over from Boeing Vertol) as the Twin Jet; the same organisation (now American Eurocopter) also used the title Super Five for a BO 105CBS with improved rotor aerofoils, and Lift Ship for the BO 105LS 'hot and high' variant built by MBB Helicopter Canada. Giraffe was the name given by MBB (along with an appropriate colour scheme) to a BO 105 (D-HABV) fitted in 1981 with an experimental Mast Mounted Sight (MMS) on top of the rotor hub, in a similar manner to that of the better known Bell OH-58D Warrior (cf **Kiowa**). The Giraffe was converted as part of the research effort for the German Army's PAH-2 anti-tank helicopter requirement (to replace the existing BO 105P or PAH-1) which is to be met by the Eurocopter Tiger tandem-seat attack helicopter.

Executive
Slightly deceptive name used briefly by Allegheny Airlines of Washington DC in the mid-fifties for their 40-seat Martin 2-0-2 twin-engined propliners.

Exploder
It may not have been the most glamorous aeroplane in the world, but in terms of adaptability and longevity the Beech Model 18 is a true classic. No less than 9,388 copies of the famous Twin Beech were built

between 1937 and 1969, and among the nicknames picked up over the years are *Wichita Wobbler*, and (in the RAF at least) *Twin Harvard*. *Exploder* was a pun on Expeditor, the generic military title introduced with the USAAF's C-45F model in 1943 in an attempt to rationalise a plethora of existing service titles: the US Navy's JRB transport had been named Voyager, the USAAF's photo-recce F-2 the Discoverer, the AT-7 trainer the Navigator, and the AT-11 trainer with a bomb bay and glazed nose was the Kansan (not Kansas, as sometimes stated). The Navy counterparts to the AT-7 and AT-11, the SNB-2 and SNB-1 respectively, were sometimes nicknamed *Sneebs*. Post-war, the Model 18 was upgraded by Beech as the Super 18, and the name Super-Liner appeared briefly in 1964 for a high-density ten-passenger version of the H18 executive aircraft.

The Model 18 must also hold some kind of record for the number of conversions applied to it. In 1964 Volpar Inc. of Van Nuys flew their Turbo 18 with Garrett TPE-331 engines, increased fuel tankage in a larger wing, and their tricycle undercarriage which had not only been offered by Beech themselves (in the Trigear 18) from 1963, but which also featured in many other companies' conversion packages. At least 25 Turbo 18s were completed, plus about 23 Turboliners, with uprated TPE-331s and a 81½in stretch. An interesting but unbuilt Volpar project was the Mini-Tanker, intended for inflight refuelling of helicopters and light-attack aircraft.

Volpar shared Turboliner 'production' with Hamilton Aviation of Tucson, whose Little Liner was a simple but highly effective clean-up of the Model D18S with a new nosecone and windscreen, and modified cowlings and tailwheel. In 1970 Hamilton also acquired the PT6A-powered Westwind programme, and over 40 were eventually completed; the Westwind 11STD was a stretch to seat up to seventeen passengers whilst retaining the taildragger gear, and was offered with a variety of turboprops, but only a single PT6A-powered example flew, in 1975.

Dumod's Infinite I (later Dumod I) had the Volpar nosewheel layout, redesigned wingtips and a new windscreen, and roughly 40 conversions are thought to have been made. Less successful with possibly only two examples was the Infinite II (later the Dumodliner) of 1964 with a 6ft stretch and a new centre fin in addition to the familiar twin-tail unit. In contrast, Pacific Airmotive's well-known Tradewind had a new single-fin unit, Volpar undercarriage, metal control surfaces and increased fuel; over twenty are known to have been converted. Rausch were unusual in not adopting the Volpar gear, and instead devised their own tricycle arrangement with units from North American P-51s and T-28s; their Star 250 also featured a deeper fuselage offering more headroom, a 4ft fuselage stretch, and a new nosecone.

The most radical of all developments must have been the Skyliner, originally planned by Volpar as the Centennial, but sold in 1976 to Aerocom of Reno, Nevada. Compared to the Volpar Turboliner from which it was developed, the Skyliner would have had a new wing, PT6A engines, a fuselage stretch of over 13ft, and a new single-fin tail, but it is doubtful whether even a prototype flew before the programme was abandoned.

Express
A cargo conversion of the long-bodied models of the Mitsubishi Mu-2 light transport by Dallas-based Mu-2 Modifications Inc., the **Express** features a new port-side cargo door and a crew-access door in place of one of the cockpit side windows; certification was in 1985. Broadly similar in both timescale and concept is the Cargoliner conversion, by Cavenaugh Aviation of Houston. In 1979 Mitsubishi introduced the names Marquise and Solitaire respectively for refined developments of its previously unnamed Mu-2N (long body) and Mu-2P (short body).

Extender
Formal name of the KC-10 tanker/transport version of the McDonnell Douglas DC-10-30CF commercial widebody, 60 of which were delivered to the USAF between 1981 and 1988.

Flying Eye
Apt name coined by the Press department of the German RLM for the Luftwaffe's Focke-Wulf Fw 189 reconnaissance aircraft, a twin-boom design with a heavily-glazed central fuselage 'pod'. Earlier, designer Kurt Tank had sought to maintain the Focke-Wulf tradition of bird names by unofficially calling it Eule (Owl) in recognition of its 'large eyes and head', but it was by the onomatopoeic variation *Uhu* that the popular Fw 189 became best known in service.

Fairly Rattle
Fitting pun on the name of the RAF's Fairey Battle, the single-engined light bomber that took such a mauling in the early days of WW2.

Faithful Annie
This popular nickname perfectly

reflects the fond regard in which the RAF held the Avro 652A Anson, particularly as it appeared relatively early in its long career. First ordered in 1934 as a coastal reconnaissance derivative of Imperial Airways' Avro 652 transport (two only) it was also known in its early years as the *Aggie*, *Limping Annie* (from the uneven sound of its twin A.S. Cheetah engines) and of course just plain *Annie*. Although it soon gave way in Coastal Command to types like the Hudson and Beaufort, the Anson remained in production as a trainer and transport until the 11,020th example in 1952, and the final RAF Anson C19s were not withdrawn until June 1968. Post-war, Avro reintroduced the Anson to the civil market as the Type 652A Nineteen (so named because it complied with the Brabazon XIX specification) which gave brief but valuable service with such long-gone operators as Westminster Airways, Hunting, and Railway Air Services.

Falcon (i)
Dassault originally called their new corporate twin-jet the Mystère XX to cash in on the fame of its Mystère fighters, but over the years the long and successful line has become equally well-known by the American marketing name of **Falcon**. The title originated as Fan Jet Falcon around 1963, when Pan Am's Business Jet Division took up American distribution of the original Mystère 20 (the former Mystère XX) although as one of the first of the new breed of small 'biz-jets' the term Baby Jet was also used for promotional purposes; the smaller and later Mystère/Falcon 10 was initially referred to in the US as the Minifalcon. The 10 and 20 were later upgraded as the 100 and 200 respectively, and other models have

included the Mystère/Falcon 50 tri-jet, and the widebody 900 tri-jet and 2000 twin. HU-25A Guardian is the formal name of the US Coast Guard's Falcon 20G derivative for rescue and offshore surveillance, although in practice these are often referred to as just Falcons. Seven of the 41 delivered in the early eighties were later modified as HU-25Bs for the monitoring of oil pollution with the Aireye sensor system, based around a Motorola APS-131 SLAR (Side-Looking Airborne Radar) pod under the fuselage; a further nine became HU-25C Interceptors or *Night Stalkers* with FLIR (Forward-Looking Infra Red) and the AN/APG-66 radar of the F-16A fighter for interdicting drug-running aircraft. Dassault also offers maritime models of the Mystère/Falcon 200 and 50 with the more Gallic titles of Gardian 2 and 50.

Falcon (ii)
Variant of the popular Italian Partenavia Oscar 180 lightplane, built under licence by AFIC of South Africa from 1967 with the designation RSA 200.

Falke
Focke-Wulf's Kurt Tank referred to his Fw 187 **Zerstorer** as the **Falke** (Falcon) because of its exceptional climbing and diving performance, but it is doubtful whether he ever intended it as a proper name. Either way, the Fw 187 never entered service, despite British belief to the contrary, and reports in the Luftwaffe magazine *Der Adler* (The Eagle) that the 'practically irresistible' twin-engined machine had 'repeatedly proved itself a valuable addition to the German Air Force'.

Fan Jet
Gulfstream American proposal for a licence-built derivative of the VFW-

Fokker VFW-614, a short-haul airliner (what would now be termed a Regional Jet) notable for having its twin Rolls-Royce M45H turbofans mounted above the wing. The GAC 616 version would have differed from its German/Dutch counterpart in having a stretched fuselage seating 60 (up from 40), extended wings with winglets, and General Electric CF-34 engines. The project was abandoned in the later summer of 1979.

Fanjet 500
Initial name for the Cessna Model 500 Citation corporate jet, changed on the occasion of its maiden flight in September 1969. Eagle is the title of a conversion package available from Sierra Industries (formerly Astec) of Texas, which improves the range of the Citation 1 and Citation 500 models by use of a new advanced wing profile, and the US Navy's T-47A trainer (Citation S/II) is supposedly known as the *Platypus* because of its new broad radar nose.

FanStar
Mid-eighties upgrade applicable to early-model Lockheed JetStar corporate aircraft (military C-140), centred around replacing the quartet of P&W JT12D turbojets with a pair of GE CF-34 turbofans, and adding 3ft winglets to the wingtips. American Aviation of Van Nuys, California, flew the prototype conversion in late 1986, and by the end of the year claimed orders for sixteen aircraft.

Farmer's Eagle
Slightly obscure local term for the Dornier Do 28D-2 Skyservant utility transports attached to the Luftwaffe's Jabo 49 at Furstenfeldbruck ('Fursty') near Munich; *Flying Wolpertinger* refers to a small creature featured in

numerous Bavarian legends. Do 28D-5X TurboSky was the name of a Skyservant fitted with Lycoming LTP-101-600 turboprops in place of the usual piston engines in 1978, but the name was dropped when development switched to the upgraded Do 128 series.

Fat Albert
Numerous American flying machines have been given the fond nickname *Fat Albert* over the years, ranging from airships and blimps down to less obvious types like the Piper Apache. In its early days the name came from scandal-hit Hollywood actor Fatty Arbuckle, but in more recent times the continuance of the name probably owes more to one of the characters created by comedian Bill Cosby.

Fat Julie
Popular nickname for the 1928 Amiot 122, a 70ft-span single-engined French bomber biplane, with a fuselage so deep it had two floors for crew mobility.

Fat Lynx
Occasional nickname for the Westland W 30 nineteen-seat transport helicopter, using the dynamic components of the military Lynx matched to a new boxy fuselage. Just 40 W 30s were built and the single biggest customer, Pawan Hans of India, had its remaining fleet of nineteen permanently grounded after a third crash by the type in 1989.

Fee
Also known as the *Flying Commode*, the Royal Aircraft Factory FE2 pusher biplanes were the great troupers of the RFC, serving in almost every conceivable combat role, and no mean fighters in their day; indeed,

the 'Eagle of Lille', the great Max Immelmann, fell to an FE2b of 25 Squadron (18 June 1916). *Chinese Scout* was the term often applied to a string of single-seat, night-fighter conversions of FE2bs and 2ds carried out in the workshops of several Home Defence units.

Fennec
For French operations in Algeria, Sud Aviation converted no less than 245 USAF-surplus North American T-28A Trojan trainers into **Fennecs** (a kind of large-eared desert fox) with more powerful Wright radials, three-blade props, and underwing hardpoints. The scheme originated with North American as the NA-260 Nomad in 1958, but after a single conversion the development was taken over by Pacific Airmotive of Burbank, who supplied Sud with the original NA conversion and three more pattern aircraft. A further T-28A conversion was the T-28-R Nomair by Tucson-based Hamilton Aircraft; the T-28-R1 was broadly similar to the Nomad (customers included the Brazilian Navy, with six) but the commercial -R2 version incorporated a new five-seat cabin with a car-type door.

Féroce
Name used by Avions Fairey for the beautiful but unsuccessful Fantôme fighter biplane of 1935, designed to Belgian Air Force requirements by the British parent company.

Fertile Myrtle
With a typical US Navy alphabet soup designation like XTB3F-1S, it is perhaps not surprising that the new Grumman anti-submarine search prototype of 1947 was quickly nicknamed *Fertile Myrtle* (American

pronunciation required) after its large belly radome. Fortunately the production version was renamed the AF Guardian, and in service the type operated in hunter-killer pairs, comprising the AF-2W *Guppy* with the aforementioned search radar, and the armed AF-2S *Scrapper*.

Field Kitchen
Nickname of the Short S47 Triple Tractor biplane of 1912, from the excessive heat produced by the pair of Gnôme engines under a shared 16ft cowling. The Triple Tractor name derived from the fact that the rear Gnôme was used to chain-drive a pair of wing-mounted tractor props, in addition to the one in the nose driven by the forward Gnôme.

Fijju
Finnish Air Force name for the 1937-vintage Fiat G50 fighter monoplane; both **Fijju** and the G50's Italian title of **Freccia** mean 'arrow'.

Firecrest
Short-lived and unofficial title used by Blackburn for their B-48, a pugnacious-looking, single-seat torpedo fighter of which two prototypes were flown between 1947 and 1948.

Firefighter
Fire-bombing variant of the well-known Britten Norman BN-2 Islander utility twin, with four internal tanks holding a maximum load of 800 litres of water or fire-retardant. Company demonstrator G-BDHU flew in Firefighter configuration in 1976, but no sales ensued. More successful Islander developments have been the military Defender series and the stretched, three-engined BN-2A Mk 3 Trislander, which was nicknamed the

Twopenny Trident early in its career. After completing 81 Trislanders, Pilatus Britten Norman sold the rights and twelve unfinished airframes in 1982 to International Aviation Corp. of Homestead, Florida, where they were to have been completed with the name Tri-Commutair; the programme collapsed with the death of IAC's president, but in 1990 it was picked up by Audrey Promotions of Australia and offered with the slightly changed name of Tri-Commuter.

Five Ton Budgie
Once described as 'the world's biggest light aircraft', only the dumpy little Shorts Skyvan light freighter could have attracted such names as the *Five Ton Budgie* (not to be confused with the HS/BAe 748 *Budgie*), *Shoebox* (or just *Box*) and the *Flying Shed*. NASA's N430NA which uses a trapeze hook system to retrieve parachute-borne weather rocket payloads in mid-air, also rejoices in the name of the *Ugly Hooker*. The **Skyliner** was a 22-seat passenger version with extra doors and quieter four-blade props, perhaps best remembered for its service with BEA/BA in Scotland from 1973, and possibly the world's most unlikely VIP aircraft was the Skyliner Executive supplied to the King of Nepal (cf *Box*).

Flagship
Title used by American Airlines from the thirties onwards for its piston-engined airliners, chiefly the Douglas DC-3 (which American launched) DC-4, DC-6, DC-7 and Convair CV-240. The introduction of jets to the fleet in the shape of the Boeing 707 in 1959 brought about the new AstroJet name, also carried by the Boeing 727, BAC One-Eleven (cf *Bacalla*) and Convair CV-990. To emphasise the

spaciousness of the widebody Boeing 747 it was briefly named AstroLiner when it entered service with American in 1971, but later that year the airline switched its marketing efforts more into the field of customer service and brought in the current names LuxuryLiner for its widebodies and LuxuryJet for the narrow-bodies; since then LuxuryLiners have included the Boeing 747 and 767, and the McDonnell Douglas DC-10 and MD-11, whilst narrow-bodied LuxuryJets have been the Boeing 757, BAe 146, McDonnell Douglas MD-80 and Fokker 100.

Flaming Coffin
The Achilles' heel of the otherwise outstanding Airco DH 4 WW1 bomber was the positioning of the 60-gallon fuel tank in the fuselage between pilot and observer, which not only hampered communication but also proved vulnerable under fighter attack. Despite its grim nickname, the DH 4 was not altogether unpopular with its crews, and was invariably preferred to the DH 9 (cf *Ninak*) that was supposed to replace it. As well as 1,449 built in Britain, Uncle Sam turned out over three times that number, jingoistically dubbed *Liberty Planes* (although in fact the name actually came from that of its Liberty engine); more than 1,800 reached France, most ending their short lives ignominiously on the so-called 'Billion-Dollar Bonfire' after the Armistice. Nevertheless, the DH 4B (with DH 9-style cockpit arrangement) became one of the foundation stones of post-war American aviation, and among a profusion of conversions were the increased fuel Dayton-Wright Nine-Hour Cruiser, DH-4B-2 Blue Bird army trainer and three-seat Cabin

Cruiser, and the Engineering Division's DH-4B-5 Honeymoon with a separate enclosed cabin for two passengers. The record-breaking Aerial Coupé was a one-off design by Orville Wright with a new four-seat enclosed cabin completely filling the gap between the wings.

Flapping Goose

Initially known in the Kingston works as the *Sigrist Bus* after its designer, the Sopwith 1½ Strutter earned its more famous title from the unusually long and outwardly-angled cabane struts (fuselage to wing) which made it seem neither a single-bay biplane or a two-bay one. The name began informally but was later given a degree of official recognition, and a less common variation was *Strut-and-a-Halfer*. The 1½ Strutter was one of the most adaptable two-seaters of WW1, although by the summer of 1917 when it was past its prime, at least one RFC squadron (No. 45) called it the *Flapping Goose* in comparison with the opposing Albatross scouts. *Comic* was the name used in 78 Squadron for an improvised single-seat night-fighter conversion, devised in late 1917 by Capt. F.W. Honnett, flown from the former gunner's position and with the usual forward-firing Vickers replaced by either one or two wing-mounted Lewis guns.

Flash

Name given to a 1973 proposal by Fairey SA to licence-build the Saab 105XT jet trainer for the Belgian Air Force.

Flying Flea (i)

It has been said that the Frenchman Henri Mignet called his infamous Do It Yourself HM 14 the Pou de Ciel (Sky Louse) because it was small and would make people scratch their heads. In Britain the tandem-wing contraption was renamed **Flying Flea** by the Air League and was duly banned in 1936 following a spate of fatal crashes.

Flying Flea (ii)

Surely the most exciting warplane of WW2, it is perhaps ironic that the rocket-powered Messerschmitt Me 163B Komet (Comet) interceptor should share this colloquialism with the pre-war Pou. The only thing the two really had in common was a notorious lethality towards their pilots, but in the case of the Komet (initially known to British Intelligence as the Peenemunde 30) the *Flying Flea* title arose from its phenomenal initial climb rate of 16,000 ft/min. Test pilot Hannah Reitsch described flying it as 'Intoxicating ... like thundering through the skies on a cannonball', and its speed at altitude of almost 600 mph also earned it such titles as the *Devil's Sled* and *Power Egg*, the latter supposedly coined by a Berlin schoolboy.

Despite the loss en route of a German pattern aircraft, Japan managed to complete a close copy known as the Mitsubishi J8M1 Shusui (Sword Stroke) only for the first aircraft to crash on landing (like so many of its German counterparts) after its maiden flight, a little over a month before VJ Day. Japan also built a wooden glider model, the MXY8 Akikusa (Autumn Grass) for training purposes, but a proposal for a ducted fan version known as the Shuka (Autumn Fire) remained a paper project only.

Flitfire

'Count the Cubs' was the proud challenge of Piper (formely Taylor) advertisements in the thirties and it became almost a cliché to say that what Henry Ford did for the automobile, Bill Piper did for the aeroplane; the analogy was taken further with the suggestion that you could have a Cub in any colour you wished, as long as it was yellow. The original Taylor Cub flew in September 1930, and major models during its record-breaking production life have included the pre-war J-2 and J-3, the military L-4 **Grasshopper**s and the post-war PA-18 Super Cub.

In nomenclature terms the Cub was unusual in that it was named after its original and soon discarded Brownach Tiger Kitten engine. Also something of a one-off is the *Flitfire* name associated with an American scheme to raise money for the RAF Benevolent Fund by raffling a new Cub; to publicise the raffle, Piper dealers in each of the States donated a silver Cub in RAF-style markings, and at a glitzy occasion at La Guardia Airport in April 1941 each was christened Flitfire Oklahoma, Flitfire Virginia etc. to raise awareness of war-torn Britain and her saviour the Spitfire. Post-war, Canadian distributor Cub Aircraft of Ontario introduced the locally-built L-4B Prospector (not to be confused with the USAAF's L-4B **Grasshopper**) for camping and bush flying, but the most unusual variant must be the Indonesian Air Force's NU-90 Belalang (Grasshopper) in which the familiar high wing was attached to the lower fuselage, the undercarriage moved forward slightly, and a new sliding canopy fitted; the prototype (designated NU-85) first flew in April 1958, and subsequently most of the air force's L-4J Cubs were converted into Belalang primary trainers at the Husein Sastranagara Air Force Base.

Since 1975, Wag-Aero of Wisconsin has supplied kits and plans for a J-3 replica with more modern materials as the CUBy (pronounced 'Cubbie') Sport Trainer and the short-wing CUBy Acro Trainer, whilst for those wanting a pseudo-warbird the Observer recreates the wartime L-4 **Grasshopper** with more extensive cabin glazing.

Ford
Taking its formal name from the manta ray and its more commonly used nickname (shared with the A-4 Skyhawk) from its designation, the US Navy's Douglas F4D Skyray was acknowledged as one of the hottest fighters of the fifties. With 16,000lb of thrust and a large, distinctively shaped wing, the F4D was once said to climb 'like a homesick angel', a facet which led to the alternative nickname of the *Ten Minute Killer*.

Flying Forest
Like the Boeing B-52, the mighty Lockheed C-5 Galaxy is surprisingly short of real, intrinsic nicknames. Apart from the usual *Fat Albert*, camouflaged C-5s are sometimes known by the title of *Flying Forest*, and an American publication once coined the rather unkind *FRED* – Fantastic, Ridiculous, Economic Disaster.

Fork-Tailed Devil
Luftwaffe fighter pilots' name for the famous Lockheed P-38 Lightning twin-boom fighter, known tentatively by its makers for a while as the Atlanta, and only christened Lightning with the introduction of the P-38D. *Droop Snoots* were P-38Js and Ls modified with a glazed nose to house a bombardier, and *Pathfinder* was the term for P-38Ls performing a

similar role with a nose-mounted 'Mickey' radar (cf *X-Engine Airplane*).

Fort
Newspaperman Richard Williams coined one of the most evocative names in aviation when, in a *Seattle Times* photo caption, he described Boeing's forthcoming five-gun Model 299 as a veritable flying fortress. By the time the short-lived prototype flew in July 1935 Boeing had registered the title as a trademark, and although the British sniffily discarded the first word as superfluous, the name went on to become a legend. Curiously, early German intelligence reports occasionally referred to the B-17 as the Boeing Seattle, and in later years the Nazi propaganda machine was to label it the *Flying Coffin*.

Four-Legged Bug
Also dubbed the *Mammoth Water Spider*, the Hughes (ex-Kellett) XH-17 Flying Crane of 1952 was one of the most unusual rotorcraft ever flown. Standing on four stilt-like legs in order to straddle projected loads of up to 27,000lb, the XH-17's 130ft rotor turned at a mere 88rpm, powered by hot gas ducted from twin turbines and ejected at the rotor tips by fuel-driven burners; almost afterburners in fact.

Freccia
Italy's Aerfer began developing their series of attractive lightweight jet fighters in 1952, when a wooden Ambrosini Supersette (Super 7) piston-engined trainer was fitted with experimental swept surfaces as the **Freccia** (Arrow), and the following year given an 840lb thrust Marbore jet exhausting under the rear fuselage to become the Sagittario (Archer). The all-metal Sagittario 2 with a Rolls-

Royce Derwent 9 became the first Italian aircraft to fly faster than sound (albeit in a dive) on 4 December 1956, but still fell short of the NATO specification at which it was aimed. Two further developments were the 1958 Ariete (Battering Ram) with a tail-mounted Rolls-Royce Soar turbojet in addition to the Derwent, and the Leone (Lion) which would have had a Bristol Orpheus or de Havilland Gyron Junior, with a DH Spectre rocket in the tail; development was abandoned when the Italian Air Force opted for the rival Fiat G91.

Freedom Fighter
With no apparent domestic requirements for its private-venture N-156F twin-engined lightweight fighter, developed in parallel with the T-38 Talon trainer, Northrop pitched it squarely at the export market with the evocative title of **Freedom Fighter**; the name gradually disappeared, however, after the Department of Defence adopted the N-156F for export under MAP (Military Assistance Programme) auspices as the F-5 in 1962. First foreign deliveries were to Iran in 1965, and around the same time the USAF also evaluated a dozen F-5As as ground-attack aircraft in Vietnam under the Skoshi Tiger programme, Skoshi being a corruption of Sokoshi, the Japanese for 'Little'. Inevitably, the aircraft themselves became known unofficially as *Skoshi Tigers* or *Little Tigers*, and though the evaluation did not lead to a USAF order the name was perpetuated by Northrop in the improved F-5E Tiger II of 1972, incorporating 5,000lb thrust GE J85-21 engines and enlarged leading-edge root extensions amongst other improvements; the RF-5E photo-reconnaissance model was

appropriately named Tigereye, and in the Taiwanese Air Force locally-assembled F-5Es are known as the Chung Cheng (another name for the late President Chiang Kai-Shek).

The ultimate devlopment of the line was the F-20 (initially F-5G) Tigershark, so named because of its new nose profile, with a single huge GE F404 turbofan of 17,000lb thrust (over twice the power installed in the F-5A) pushing it into the Mach 2 class. Sadly, the F-20 failed to attract an order, and two of the three prototypes were lost in fatal crashes. F-5 Plus Tiger III is the name of an Israel Aircraft Industries upgrade devised for Chile (Israel itself is not an F-5 operator) with an updated cockpit and weapons system.

Apart from the *Skoshi Tiger*s, the closest the American services have come to using any F-5s as fighters has been the F-5Es and two-seat F-5Fs used as aggressor aircraft for Dissimilar Air Combat Training (DACT), simulating Warsaw Pact aircraft in tactics, operating procedures and even camouflage. Having brought many an F-15 and F-16 jockey down to earth, metaphorically speaking, over the Nellis 'Red Flag' ranges, the USAF F-5s were sometimes known as *Humiliator*s or *MiG-5*s, and *Gomershark* was the product of wishful thinking regarding the F-20 Tigershark, a 'Gomer' being an all-purpose bad guy.

Frightful
Test pilots Gerry Sayer and Michael Daunt (seconded from Gloster) are credited with this name for the Folland 43/37 (F108), a terrifying Beaufighter-sized, single-engined monoplane built purely for engine-development work. Of the dozen built

from 1940 and used on the Sabre, Hercules and Centaurus engine programmes, seven are believed to have been written-off in accidents.

Frightning
Reference to the alarming crosswind landing characteristics of the late lamented EE/BAC Lightning interceptor, brought about by its flat-sided fuselage. Excalibur was among the formal names considered for what turned out to be the RAF's last real all-British fighter.

Frog
It's green and it hops, but to really understand this popular US Marine Corps nickname you have to see a Boeing CH-46 Sea Knight helicopter from head-on. Canadian Armed Forces variants of the tandem-rotor Model 107 family are the search-and-rescue CH-113 Labrador (six acquired) and the CH-113A Voyageur transport (twelve acquired) and in the Japanese Self-Defence Forces the Kawasaki-built KV-107 is known as the Shirasagi (Snow Heron).

Furniture Van
Crews' nickname for the large and cumbersome Junkers J I (not to be confused with the J 1 *Tin Donkey*) metal-skinned recce biplane of late 1917.

Gajraj
Broadly comparable to the Lockheed C-141 StarLifter, the Soviet/CIS Ilyushin Il-76 freighter (NATO – Candid) is also operated in its Il-76MD form by the Indian Air Force as the **Gajraj** (King Elephant). In 1989 Iraq unveiled a local AEW adaptation of the Il-76MD known as the Baghdad 1, with a licence-built Thompson CSF Tigre surveillance radar mounted inverted in the position

usually occupied by the I1-76's rear-loading doors. Development was later switched to a more conventional AEW configuration with a dorsal-mounted rotodome, in which form the aircraft was renamed Adnan, after a formal Iraqi defence minister; the exact status of the programme is uncertain, but at least one Adnan is thought to be in service with the Iraqi Air Force. Iraqi AEW conversions should not be confused with the Soviet/CIS Beriev A-50 (NATO – Mainstay) AEW derivative of the I1-76, which was initially dubbed *SUWACS* (i.e. Soviet Union AWACS) by sections of the Western aviation press.

Gas Pipe Aeroplane
Name originating at the British & Colonial Aeroplane Co. ('Bristol') for the Royal Aircraft Factory's BE10, an offshoot of the well-known BE2c with a then-novel steel tube structure; no BE10s were completed by either organisation.

Gawron
A Polish development of the Soviet Union's long-serving Yak-12 (NATO – Creek) high-wing monoplane, the PZL-101 **Gawron** (Rook) featured a simplified structure affording a greater payload. A total of 330 were built, mainly for agricultural work, between 1958 and 1970.

Gelatik
Gelatik (Rice Bird) was the name of the 39 Polish PZL-104 Wilga (Thrush) STOL lightplanes built under licence by the Lipnur organisation in Indonesia from 1965, and powered by Continental 0-470 engines.

Georgia
TBV Georgia was the title of a

proposed US Navy torpedo bomber development of the USAAF's much-maligned Vultee A-35 Vengeance WW2 dive-bomber.

Gestapo Aircraft
Slightly obscure Royal Navy term for a Japanese aircraft used to shepherd kamikaze suicide aircraft to their targets in WW2. The aircraft concerned would usually be equipped with radar and flown by an experienced crew, and were used to overcome the navigational shortcomings of the hastily trained kamikaze ('divine wind') pilots.

Giant
Generic term for the remarkable German R Class (Riesenflugzeug or Giant Aircraft) heavy bombers of WW1, embracing such types as the AEG RI, DFW RI, Linke-Hofmann RI and RII, Siemens-Schuckert RI to RVII, and the Zeppelin Staaken RVI. For comparison, the 138ft span of the Zeppelin was just 3ft short of a WW2 Boeing B-29 Superfortress.

Gillette Falcon
Taking its nickname from the well-known brand of razor blades, a single Miles M3E Falcon Six cabin monoplane (L9705) was modified to flight-test a wooden version of the thin wing intended for the cancelled M52 supersonic research aircraft.

Gina
Widely used nickname for the Aeritalia G91R light ground-attack fighter. The G91T two-seat trainer variant is fondly known as the *Virus* in its native Italy.

Ginny
Affectionate term for the Vickers Virginia, the slow and stately heavy

bomber biplane that so epitomised the
leisurely days of the RAF between the
wars. Vickers initially planned to call
it the Vulcan.

Gnome Sandwich
Nothing to do with undersized
aviators, this memorable nickname
was applied to the 1911 Short
Tandem-Twin biplane because of the
close proximity to the pilot of its fore-
and-aft 50hp Gnôme rotary engines.
With the pusher prop a mere 10in
behind the pilot's head, the air current
in the open cockpit was reputedly
strong enough to pluck the hairs from
a fur coat, and combined with the
access opening in the cockpit floor this
led to the alternative nickname of the
Vacuum Cleaner.

Goat
Along with *Duckbutt* and the almost
mandatory *Dumbo*, *Goat* was a
nickname for the twin-prop Grumman
SA-16 Albatross amphibian, perhaps
best remembered for its air-sea rescue
work in the Korean War. At the time
of its maiden flight in October 1947
the type was formally known as the
XJR2F-1 Pelican.

Goldfish Bowl
In the most familiar of its many guises
the Bell Model 47 was the archetypal
'bubble and boom' type of helicopter,
and its appearance is known to
millions through the M.A.S.H. TV
series. Whilst there was no overall
name for the Model 47 as a whole,
Sioux became commonplace through
its use by the US Army and Air Force
(with the H-13 designation) and by the
British Army and RAF for their
Westland-built Model 47G-3B-1s;
Trooper had been adopted in 1959 as
a commercial or marketing name for
the 47G, but the British opted for the

Sioux name to follow on from the
Skeeter and Scout helicopters. Sioux
Scout was the name of the 1963 Model
207, a purely experimental gunship
version with tandem seating, stub-
wings and nose-mounted gun turret.
    The deluxe Model 47H was given
the slightly odd title of Bellairus, and
the 47J series with enclosed cabin and
covered-in tailboom was the **Ranger**;
Super Ranger was the uprated 47J-3
built by Agusta in Italy. Another
licensee was Kawasaki of Japan,
where the 47G was given the JASDF
name of Hibari (Skylark) in 1964; in
Vietnam the Australians adopted the
nickname *Possum* for the 47G, whilst
ex-RAF machines in the Royal
Malaysian Air Force were known as
the *Belap Siap (Unfinished)*.
    Among the numerous conversions
based on the Model 47 since it became
the world's first commerically
certificated helicopter in March 1946
were the Texas Helicopters/Aerodyne
M-79T Hornet armed trainer and
single-seat M-74 Wasp crop-sprayer,
both of which appeared in the mid-
eighties, but the most successful must
be the series of El Tomcat single-seat
crop-sprayers, certificated by
Continental Copters of Fort Worth in
1959 and available until 1981.

Goliath
Original name within Junkers for the
huge Ju 322 Mammut (Mammoth)
glider, built to the same requirement
as the better known Me 321 Gigant (cf
Sticking Plaster Bomber). Also
known to British Intelligence as the
Merseburg (where it was built) the Ju
322 was found to be hopelessly
unstable and was quickly abandoned.

Gomhouria
A relatively straightforward variant of
the Luftwaffe's Bucker Bu 181

Bestmann trainer/communications aircraft, the **Gomhouria** (Republic) remained in production at the Heliopolis Aircraft Works, Egypt until as late as 1979, by which time over 300 had been built.

Gonad
Slang for the Australian GAF Nomad light STOL transport, first flown in 1971. Whilst Nomad remained the formal type name, variants were also given such self-explanatory titles as Searchmaster, Floatmaster, Medicmaster, Surveymaster, **Cargomaster**, Commuterliner, and the military Missionmaster.

Gonzo
Fitted with an elongated radar nose for navigation training, the Canadian Air Force's quartet of Boeing Canada (ex-DHC) CT-142s are popularly known by the name *Gonzo*, after the beaky character from the *Muppets* TV show. The CT-142 and transport CC-142 are militarised variants of the Dash-8 twin-turboprop commuter, and since 1988 the company has also offered various armed maritime models under the generic name of Triton.

Gooney Bird
There are those who talk of *Gooneys* and those to whom the name Dakota springs more readily to mind, but few could disagree that the Douglas DC-3/C-47 series is by far the most significant transport aircraft the world has yet seen. Worldwide production is estimated at 12,654, and at the start of the nineties there were still over 600 in civil and military service worldwide.

The first of the line, the DST (Douglas Sleeper Transport), was promoted around the time of its maiden flight on 17 December 1935 as the Sky Sleeper, although it did not really amount to a formal name; various internal layouts were offered with the names of Skylounge with fourteen convertible seat/beds, and the more private two-person Sky Room (nicknamed the Honeymoon Hut) in the forward cabin. Just 38 DSTs were built, but orders for the dayplane DC-3 development flooded in and by 1939 the type was carrying no less than 90 per cent of world airline traffic.

Principal wartime versions were the C-53 Skytrooper personnel transport (almost 400 built) and the slightly later C-47 Skytrain freighter/transport (roughly 9,200 built), and it was in US military service that the famous *Gooney Bird* name first appeared, after the awkward-looking species of albatross; C-47s air-dropping supplies were also occasionally referred to as *Biscuit Bombers*. Among other wartime operators were the US Navy (the R4D), the Soviet Union (where it was licence-built as the PS-84 and Li 2; NATO – Cab), Japan (the L2D, built by Showa and Nakajima, reporting name Tabby) and of course the RAF and Commonwealth air forces, who took delivery of almost 1,900 and introduced the Dakota name. Some sources claim the name was chosen because it is almost an acronym for 'Douglas Aircraft Company Transport Aircraft', but this seems unlikely.

The USAF's post-war SC-47 search-and-rescue variant inevitably became another *Dumbo*, and as late as the Vietnam War the EC-47 variant of the faithful old *Gooney* was pressed into service dispensing propaganda via leaflet and loudspeaker as the *Bullshit Bomber* or *Sister Gabby*. Perhaps better known from the SEA campaign though was the AC-47 Spooky gunship, originally to have been the FC-47 until the fighter community objected. An American journalist likened the tracer and staccato roar from the trio of 6,000rpm miniguns to a fire-spitting dragon, although other sources suggest the analogy originated with friendly South Vietnamese forces; either way, the AC-47 became referred to in-theatre as *Puff, the Magic Dragon* (aka *Puffship* or *Dragonship*) after the 1964 pop song by Peter, Paul and Mary. There is also evidence to suggest the name was taken literally by many of the notoriously superstitious Vietcong, who believed the night-flying AC-47 was a real live Yankee dragon.

One of the best known civil Dakota conversions was BEA's 38-strong Pionair class (individual aircraft were named after aviation personalities) which dispensed with the radio operator and introduced British instrumentation, an improved 38-seat cabin and integral airstairs; a further ten freighter models retained the familiar double-doors however and were known as Pionair Leopards. Prior to the introduction of the first full conversion (G-ALYF) in January 1951 there were three with just the cockpit mod and the hybrid title of *Piota*. Eight Pionairs were later sold to Welsh airline Cambrian, who in October 1967 introduced the Super Pionair, also dubbed *The Dream of Elwin* after MD Bill Elwin; the new luxury interior was considered somewhat incongruous in such a geriatric airframe however, and only one (G-AHCZ) was ever converted.

In the early fifties Pan Am/Panagra developed the Hi Per DC-3 with higher-powered P&W R-2000 motors and new props for operation from high-level airports, and other relatively minor upgrades to engines,

windows, interiors etc. led to such airline names as **Skyliner** (New Zealand National Airways Corp.), Viewmaster (Airlines of New Zealand) and Dakmaster (the British independent Transair). More generally, the ageing fleet of DC-3s/ Dakotas worldwide has been described in such fanciful terms as the *Dowager Duchess, Old Methuselah, Old Bucket Seats, The Placid Plodder, Duck* and *Dizzy Three*.

The first turboprop conversion was a single Dakota IV (KJ839) fitted by Armstrong-Siddeley with Mamba ASMa 3s in 1949, though perhaps better remembered from the same period are the trio of Rolls-Royce Dart Dakotas, two of which served briefly as freighters with BEA. Conroy Aircraft of California also fitted Darts to produce their Turbo Three of 1969 and Super Turbo Three (converted from a post-war Super DC-3S) of 1974; both were one-offs but the former (originally a C-53 Skytrooper) re-emerged in 1977 as the sole Specialised Aircraft Tri Turbo-3 with three P&WC PT6A-45 turboprops, exhibited at Farnborough the following year in maritime patrol guise. The final attempt to market a Dart conversion appears to have been the TAMCO Turbo Commuter project of the late seventies.

In the summer of 1982 United Aircraft Corp. (USAC) flew the Turbo Express DC-3 with a pair of PT6A-45R engines and compensatory 40in fuselage plug ahead of the wing. Again, only one (N300TX) was converted, but Wisconsin-based Basler later purchased the technology and developed their successful Turbo 67R or BT-67 (PT6A-67R engines), since exported to El Salvador (two as AC-47 gunships), Colombia, Bolivia and Guatemala, amongst others. The rival DC-3-65TP **Cargomaster** conversion first flown in 1986 by Schafer/AMI (Aero Modifications Inc.) features the PT6A-65AR, and customers have included Professional Aviation Services of South Africa, who use kits to produce the DC-3AMI or Jet Prop DC-3: broadly similar also is the South African Air Force's C-47TP Super Dakota conversion programme. The SAAF formally uses the Dakota name for all its variants (C-47s included) and when they were forced to replace their ancient Avro Shackleton maritime aircraft with prehistoric *Daks* in 1984, the latter became known as *Dakletons*.

Gormless
Name coined by test pilot Bill 'Otto' Waterton for the Gloster E1/44, a rather uninspiring experimental jet fighter which Glosters had optimistically (and unofficially) labelled Ace.

Goshawk
Goshawk is frequently but erroneously quoted as a provisional name for the Gloster TC33 bomber-transport prototype of 1932, and the error is believed to stem from the fact that a version was once planned with Rolls-Royce Goshawk engines as an alternative to the Kestrels actually fitted. Better known at the time by its specification number of C16/28, the TC33 was Gloster's only four-engined type and also its biggest aircraft, although the Javelin jet fighter was heavier.

Gosling
Appropriate British name for the WW2 Grumman G-44 Widgeon amphibian, in many respects a scaled-down G-21 Goose. The G-44A was licence-built in post-war France as the

SCAN-30, and in the late-fifties about fifteen of these were converted in the US by Pacific Aircraft Engineering Corp. (PACE) into Gannets, with new all-metal wings, updated systems, and R-680E radials in place of the usual inline engines. *Petulant Porpoise* was the name of a Widgeon used by Edo Aircraft to test various hull designs.

Graf Zeppelin
It was said of America's first jet night-fighter, the Northrop F-89 Scorpion, that everything was big except the engines, and the inordinately large mainwheels also earned it the alternative nickname of the *Stanley Steamer* after the pioneer road vehicle. *FOD Vacuum* stemmed from the low-set intakes which produced an unfortunate tendency to ingest debris and cause FOD (Foreign Object Damage) to the engines.

Grand
Formally, it was the Sikorsky S-21 Russkii Vitiaz (Russian Knight), but even the more familiar nickname of *Bolshoi* (*Grand*) barely seems adequate for the aeroplane that in 1913 (less than ten years after the Wright Bros) was not only the world's biggest with a span of 92ft (roughly Boeing 737-sized) but also pioneered the use of four engines, a multi-wheel undercarriage, and enclosed cabin. Built by the Russo-Baltic Wagon Works (R-BVZ), a builder of streetcars, the S-21 was also dubbed *Big Baltic* and (with the cabin also in mind) *Flying Tramcar*; the latter title also passed on to the WW1 bomber derivative, the Ilya Mouromets.

Grandmother
Testifying to its longevity in Soviet service, the MiG-15UTI (NATO –

Midget) trainer was widely known as both the *Matushka* (*Mother*) and even *Baboushka* (*Grandmother*). The original MiG-15 fighter was reportedly known as *The Soldier's Aircraft* by groundcrew impressed by its ease of maintenance, but this would seem more than a little contrived and artificial. The MiG-15 and improved MiG-15bis were sometimes referred to as the Jaguar and Eagle respectively in the Hungarian Air Force, but again the precise status of these names is uncertain.

Grand Piano
Russia pioneered wooden-skinned aircraft in 1912, but perhaps held on to the idea too long with the WW2 LaGG-3 fighter, covered in highly polished resin-impregnated wood that bore more than a passing resemblance to the Cherrywood used for pianos. Possessed also of a vicious habit of stalling on the approach, LaGG-3 pilots christened it the *Mortician's Mate*, and changed the LaGG acronym (Lavochkin, Gorbunov & Gudkov) to *Lakirovanny Garantirovanny Grob – Varnished Guaranteed Coffin*. In view of such black humour, it could be assumed there was just a touch of sarcasm in *Wooden Wonder* . . .

Grasshopper
Unusal among aircraft nomenclature, this was the formal name introduced in 1942 (along with the new L for Liaison designation) for the USAAF's new breed of lightweight observation and liaison types, formerly designated within the O (Observation) category and loosely known by the names of their civilian counterparts. Four prinicipal types became **Grasshopper**s, namely the Taylorcraft L-2 (ex-O-57/ civil model D), Aeronca L-3 (ex-O-58/

civil model 65TC Defender), Piper L-4 (ex-O-59/civil Cub) and Interstate L-6 (ex-O-63/civil model S-1B Cadet). In the case of L-4 the well-established title of Cub continued to hold sway in everyday use, and all four (along with other non-Grasshopper liaison types such as the Stinson L-5 Sentinel) were colloquially dubbed *Flying Jeeps*.

Flying Greenhouse
Nickname of the one-off Henderson HSF 1 twin-boom pusher monoplane (G-EBVF) of 1929, when fitted with a long and heavily glazed hood over its six-seat cabin. Later of course, 'Greenhouse' became a common description for the long canopies of types like the T-6 Texan/Harvard, Fairey Battle, etc. although it was in use as early as 1916 in relation to the unpopular and quickly discarded structure originally fitted to the Royal Aircraft Factory SE5a.

Grey Goose
Initial title of the 1937 Grumman G-21 Goose civil amphibian.

Griffon
Name reportedly under consideration in 1988 for a projected RAF utility version of the EH Industries (Agusta Westland) EH101 helicopter, although no mention of the name was made when the order was finally confirmed nearly five years later. The Royal Navy version of the three-engined EH101 is of course the Merlin, and the commercial transport model is the Heliliner. In 1993 Canada chose the titles CH-148 Petrel and CH-149 Chimo respectively for the 28 naval and fifteen SAR (Search-And-Rescue) EH101s it had on order, but the entire Canadian programme was cancelled following a change of government later that year.

Grim Reaper
The sensation of the 1936 Paris Salon was the then unflown Fokker GI twin-engined fighter offering up to nine machine-guns, which immediately earned it such evocative sobriquets as the *Grim Reaper* and *The Mower*; in the hype it was apparently forgotten that Britain was already flying its eight-gun **Hurricane** and Spitfire single-seaters.

Gripper, The
Famous for its pioneering Smiths Autoland system, the Hawker Siddeley Trident jet airliner also displayed a marked reluctance to depart from terra firma on occasions, earning the fond nickname of *The Gripper* from its BEA crews. *Tripod* and *Rodent* were merely plays on the Trident name.

Grizzly
Name used unofficially by Beech for their twin-engined XA-38 attack aircraft, built around a nose-mounted 75mm cannon. Also referred to occasionally as the Destroyer, two XA-38s were flown in 1944 and 1945.

Guardsvan
Pun on the name of the Vickers Vanguard turboprop airliner, along with *Mudguard*. Overtaken by the new generation of short-haul jets in the early sixties, just 43 were sold, and both original customers later initiated conversions to exploit the type's cargo potential. TCA/Air Canada were first with their one-off Cargoliner (CF-TKK) in 1966, featuring a new freight door and rollers on the main deck, but perhaps better known was BEA's broadly similar Merchantman, nine conversions being carried out in a joint programme with Aviation Traders.

Guepard
Changed when Nord took over the project from SFECMAS, **Guepard** (Cheetah) was the oringinal name of the Nord 1500-002 Griffon II, the French turbo-ramjet-powered research aircraft first flown (as the jet-only 1500-001 Griffon I) in September 1955.

Gunbus
Though widely associated with the 1914 Vickers FB5, the first real British fighter, *Gunbus* is in fact a somewhat spurious title believed to have originated in the pre-war Press. Certainly the title was unofficial, but even as a nickname it was very rarely used in WW1, those involved with the FB5 usually referring to it simply as the Vickers pusher, gun-carrier or just the fighting biplane; the so-called *Streamlined Gunbus* was the cleaned-up Vickers FB9 of 1915. The **Gunbus** name also came to be used in a more general sense to identify the two-seat pusher layout, as exemplified by the Sopwith Type 806, and the 36 unused machines purchased by the Admiralty in 1915 from the W. Starling Burgess Co. and Curtiss in America.

Guzomee Bird
Latter-day RAF expression (pronounced 'guz-*oh*-mee') for any aircraft that brings you home from a foreign posting, especially if it is for the last time.

Hadrian
British name for the Waco CG-4A fifteen-seat assault glider of WW2, known in its American homeland as the Haig.

Halcon
T-36 Halcon (Hawk) is the Chilean Air Force title for the Spanish CASA

C-101 Aviojet jet trainer, progressive
licence-manufacture of which was
undertaken locally by ENAER in the
early eighties. A higher-powered
attack version also serves in Chile as
the A-36.

Hallybag
Standard RAF-speak for the generally
unsung Handley Page Halifax heavy
bomber of WW2. Despite the old
Yorkshire saying about 'Hull, Hell
and Halifax', few crews stationed in
the county with No. 4 Group would
have swapped their *Hallys*, which
(among other virtues) usually got
them home from a raid before
accompanying Lancasters regained
their Lincolnshire bases. As with
many other types built by more than
one company (e.g. the *Stringbag*)
experienced crews could often tell the
origin of a particular Halifax by the
way it handled, and those from the
London Aircraft Production Group
were known to the cognoscenti as the
London Bus. A total of 6,176 Halifaxes
were built, and in addition to
numerous post-war civilian freighter
conversions, twelve Halifax C8
transports were converted by Shorts
into ten-passenger Halton airliners for
BOAC, serving as interim types on the
Lagos and Karachi services until 1948.
One Halton (G- AHDM 'Falmouth')
met an undignified end in 1950 when,
with heavily modified nose and wings,
Brabazon nosewheel and a somewhat
bizarre tail unit, it was the non-flying
representation of the fictional
Reindeer airliner in the film version of
Nevil Shute's *No Highway*.

Handley Page
Long-obsolete term once found in
many English dictionaries to denote
any kind of 'large aeroplane'. Handley
Page themselves first gained a

reputation for outsize aircraft with
their WW1 O/100, O/400 and V/1500
bombers (cf *Bloody Paralyser*).

Hangar Queen
Along with its alfresco equivalent
Ramp Rooster, this term originated in
America and is applied to any aircraft
with persistent serviceability
problems. Although mainly applicable
to individual machines, both terms
have also been used in the wider sense
for an aircraft type in general.

Hannoverana
Generic Royal Flying Corps term for
the series of German WW1 Hannover
two-seaters (notably the CL II and CL
III) characterised by their unusual
biplane tail units.

Harmattan
Harmattan (a dry, West African wind)
was the title first applied to the 1953
Dassault MD 453 Mystère de Nuit, a
one-off tandem two-seat all-weather
variant of the standard MD452
Mystère II day fighter, differing in its
side-mounted intakes and 'solid' radar
nose.

Harpoon
Name proposed by designer Godfrey
Lee for what became the last of the V-
Bombers, the Handley Page Victor.
The Victor had a brief and somewhat
unlikely film role in the sixties as the
'Giles-Thompson T/GT/3 Iron Maiden
supersonic jetliner', although the real
eponymous star of the Anglo-
Amalgamated Michael Craig/Anne
Helm film was a showman's steam
traction engine.

Harry Tate
RFC rhyming slang name for the
Royal Aircraft Factory RE8 Corps
reconnaissance biplane, after the

contemporary music hall comedian famous for his routine about a flying machine with the aerodynamics of a penguin.

Hartebeest
South African Air Force version of the Hawker Audax Army Co-op biplane, four of which were built in the UK and a further 65 by the SAAF depot at Roberts Heights, Pretoria (a hartebeest is a kind of African antelope and various spellings exist, including hartbee and hartebeeste). The original Audax, one of a string of types descended from the famous Hart bomber, also served with the Iraqi Air Force as the Nisr.

Hawk
Name originally used for the Boulton & Paul P3 Bobolink (a kind of reed-bird) British fighter prototype of 1918.

Hejja
Although the Italian Regia Aeronautica rejected the home-grown Reggiane Re 2000 Falco 1 fighter, almost 200 were built by the Hungarian MAVAG organisation as the **Hejja** (Hawk) and used mainly on the Russian Front in WW2.

Heli-Porter
Slightly curious early name for the Fairchild-built model of the antiquated-looking but highly effective Pilatus PC-6 Turbo Porter STOL utility aircraft. Fairchild also developed an armed counter-insurgency model as the AU-23A **Peacemaker**, which was evaluated by the USAF and sold to Thailand (33 copies), but the pair of unarmed aircraft acquired in 1979 for US Army use in Berlin under the designation UV-20A Chiricahua came from the original Swiss production line.

Hen Turkey
Spanish Nationalist term during the civil war for the lumbering Heinkel He 46C parasol reconnaissance aircraft.

Hepcat Kitten
Initial title of the little-known Grumman G-63 Kitten single-engined lightplane of 1944. The original Kitten I taildragger was followed in 1946 by the tricycle-gear G-73 Kitten II, but neither model went into large-scale production.

Herk
If the *Air Force One* machines are the most frequently seen individual aircraft, then the most commonly seen *type* must be the Lockheed C-130 Hercules tactical transport, so often in the news associated with humanitarian operations. First flown in August 1954, the *Herk* or *Herky Bird* is as seemingly irreplaceable as the DC-3, and looks set to remain in production for some time to come, perhaps in the form of the advanced C-130J Hercules II. Diminutives apart, the most widely used nickname in both the US and UK is currently *Fat Albert* and its application to the type is often attributed to the US Navy's Blue Angels flight demonstration team, whose KC-130F support aircraft once bore the proud legend *Fat Albert Airlines* above the crew door. Less common are *Flying Dumptruck*, *Flying Jeep*, and the memorable *Screaming Green & Brown Trash Hauler* from the days of south-east Asia camouflage; more conservative were the RAF's *Brownie* and *Chocolate Lorry* from their original Middle East colours. Early C-130As lacking the now-familiar nose radome were 'Roman Nose' Hercules. In the Middle East the C-130 is supposedly nicknamed the *Air Camel*, and

Argentinian KC-130H tankers used to 'suckle' fighters during the Falklands War were known as the *Chancha* or *Mother Sow*. Airlifter was reportedly the formal name chosen for the long-range logistical C-130E in 1961, but whatever the intended status of the name it proved short-lived. Heavily-armed Spectre gunship models have included the AC-130A, E, H, and the latest AC-130U (*U-Boat*) Spectre II converted by Rockwell, with a sophisticated radar and infra-red sensor suite plus port-side-firing 25mm and 40mm cannons and 105mm howitzer.

In Vietnam, conventional transport C-130s were often called *Klong Bird*s after the canals of south-east Asia (particularly Bangkok) or *Flying Cow*s when hauling up to 6,000lb of fuel in cargo-hold bladders or 'udders'. *Crown* and *King* were successive titles (originating as callsigns) associated with the HC-130H and helicopter-refuelling HC-130Ps used as control aircraft on combat rescue missions; the latest air force tanker version is the HC-130N (Combat Shadow). *Blind Bat* was the unofficial title (again originating as a callsign) for the C-130As using flares to illuminate sections of the Ho Chi Minh Trail for night-time operations, and ten C-130E-IIs (later EC-130Es) used as Airborne Battlefield Command Control Centres were irreverently known as *Head Shed*s.

Blackbird was an early nickname for the C-130E-I (later MC-130E) used for low-level night-time infiltration/exfiltration of covert forces and recovery of downed aircrew, although in recent years this has become better known by the Combat Talon codename, which refers to the role and configuration of the aircraft rather than to the actual aircraft itself.

Early Combat Talon sub-variants comprised the MC-130E-Y (Yank), MC-130E-C (Clamp) and MC-130E-S (Swap) models named according to the particular equipment fitted, and the latest variant is the bad weather MC-130H (Combat Talon II). Other mission-adapted models and their associated codenames include the EC-130H (Compass Call) jammer, the EC-130E (Coronet Solo II) electronic surveillance aircraft, and the EC-130 (Volant Solo II) psychological warfare broadcasting aircraft (two EC-130E sub-versions, Comfy Levi and Rivet Rider, plus Senior Hunter support aircraft). Aircraft in Rivet Rider configuration are nicknamed the *Flying Speed Brake* by their crews after the huge blade aerial in front of the fin and the underwing 'Pizza Cutter' fairings.

Heron
Name formally adopted on 12 April 1994 for the five German Grob G115-D2 piston-engined trainers newly acquired by Shorts and operated by them for Royal Navy flying grading at Roborough, Plymouth.

Hertfordshire
Militarised version of the pre-war de Havilland DH 95 Flamingo twin-engined airliner, intended for the RAF; just one true **Hertfordshire** was delivered (an order for 30 was cancelled) but seven DH 95s were either impressed or diverted from civil production, and served with the RAF and RN under their original civil name of Flamingo.

Highlander
Formed by world land-speed record-holder Richard Noble, Isle of Wight-based ARV Aviation flew its first Super2 two-seat lightplane in March

1985, but financial difficulties and problems with the Hewland two-stroke engine dogged the venture, and only around 30 were built. The programme was taken over first by Island Aircraft and then Aviation (Scotland) who in 1993 went into partnership with Uvan Invest to form ASL Sweden; with a new Bombardier Rotax 912A engine the aircraft was relaunched as the Opus 280, and a home-build kitplane version was also announced as the **Highlander**.

Hirundo
Initial title (meaning 'swallow') for Italy's attractive Agusta A109 helicopter, dropped around the time of service-entry in 1976 because it was considered difficult to understand in an international marketplace.

Hiss
With its original HSS-1 designation it was inevitable that the US Navy version of the Sikorsky S-58 helicopter, the Seabat, would become known as the *Hiss*, but true to form the Marine Corps had their own way of doing things; their version was officially the HUS Seahorse, but to the leathernecks it was usually the *Huss*, reflecting not only the formal name and designation, but also its reputation as a tough, reliable old warhorse. The basic Sikorsky S-58 design was also used by the US Army as the H-34 Choctaw, and licence-built by Westland in the UK as the turbine-powered Wessex. Among British nicknames are *Camel* for the Royal Navy's Wessex HAS3 with its dorsal radome, and the well-known *Jungly* HU5s used for Royal Marines transportation tasks.

Hog
Early models of the unswept Republic

F-84 Thunderjet fighter were notoriously underpowered and sluggish on take-off; pilots branded it the *Hog* or *Groundhog*, and many swore there was a secret 'Dirt Sniffer' which would only allow lift-off at the last possible moment, when it sensed the end of the runway approaching fast. The F-84E and G models were an improvement, but by then the name had become a generic term for Republic jet fighters (cf *Super Hog*, *Thud* and *Warthog*).

Hookey Tookey
Early and popular name for the US Navy's Kaman H-2 Seasprite helicopter, derived from its pre-1962 designation of HU2K-1. Seasprites engaged on search-and-rescue missions around the Gulf of Tonkin during the Vietnam War were known as *Clementines* or, less commonly, *Angels*. Perhaps the least remembered variant of the adaptable H-2 series was the Tomahawk, a small batch of which were built for US Army evaluation in the early-sixties; an armed version had four flexible-mounted machine-guns in a redesigned nose, and another was converted into a high-speed compound helicopter with fixed wings and jet-augmentation.

Hoover
US Navy nickname for the Lockheed S-3 Viking anti-submarine aircraft, after the sound made on deck by its twin General Electric TF34 turbofans.

Horace
Popular term within the Royal Naval Air Service for the French Farman F40 biplane, derived from the fact that it was a collaborative effort by both Farman brothers, Henri and Maurice.

Horizon
The one-off Sud Caravelle 10A
Horizon or Super A (c/n 63) was an
improved version of the classic French
airliner evolved to meet a TWA
requirement ultimately filled by the
Boeing 727; improvements included
GE CJ805-23 turbofans in place of
the usual RR Avons, repositioned
cabin windows, and an acorn fairing at
the fin/tailplane intersection. A
similar variant with P&W JT8D
engines was originally to have been
called the Horizon B, but the name
was dropped when launch customer
Finnair chose to name it Caravelle
Super A.

Hornet (i)
Original name of the classic inter-war
Hawker Fury interceptor biplane, the
RAF's first 200 mph fighter; the
change of name was to comply with
the 1927 policy of having land-based
fighter names begin with the letter 'F'
(see Appendix 3), and among other
names considered were Flash, Flicker,
and Foil. The Royal Navy's broad
counterpart, the Nimrod, was initially
known as the Norn, a contraction of
'Naval Hornet'.

Hornet (ii)
Widely used but unofficial name for
the Messerschmitt Me 410 twin-
engined fighter-bomber, derived from
its use by the much-publicised II/
ZG26 'Hornissengeschwader'.

Hornet (iii)
Not to be confused with the more
famous Hiller Hornet of 1950, the
Hiller (Rogerson) RH-1100M **Hornet**
was a low-cost gunship development
of the long-established FH-1100C
utility helicopter, and was flown in
prototype form in April 1985.

Hornisse
Fa 226 **Hornisse** (Hornet) was the
original title of the German Focke-
Achgelis twin-rotor (mounted on
outriggers) helicopter project; prior to
its first free flight in August 1940 the
project came under military aegis and
was renamed Fa 223 Drache
(Dragon).

Hoverfly
Hoverfly 1 was the British title for the
unnamed Vought Sikorsky R-4 (VS-
316A) helicopter, some 52 of which
were supplied to the RAF and RN
under lend-lease from 1943. Although
the boxy, fabric-covered fuselage of
the R-4 gave way to a more
streamlined appearance with the later
R-6 the all-important transmission and
rotor system remained the same,
hence the Sikorsky designation of VS-
316B and British title Hoverfly 2 for
what appears outwardly a completely
different helicopter. The RAF and
RN received a total of 26 Hoverfly 2s
between them.

Huey
As a nickname for the classic Bell UH-
1 utility helicopter, *Huey* (from the
early HU-1 designation) has become
so widely accepted that perhaps the
real 'alternative' name is its official US
Army title of Iroquois. The
characteristic slapping noise of its
twin-bladed rotor soon became an
everyday sound in parts of Vietnam,
and in the airmobile operations which
first made it famous there were two
major breeds of *Huey*: *Slicks* were the
basic troop-carriers with just M-60
machine-guns firing from the doors for
self-defence, whilst *Hogs* were those
adapted for the external carriage of
heavier 'artillery' type ordnance;
depending on the particular
combination of weaponry carried, the

latter were also known as *Heavy Hogs*, *Frogs* and (much rarer) *Thumpers*. The slightly less common name of *Cobra* for the gunships is believed to have originated with the 114th Airmobile and was later formally adopted for the two-seat AH-1 (cf *Snake*). The Royal Australian Air Force also operated UH-1Ds and UH-1Hs as *Bushranger* gunships in Vietnam. A number of US Army conversions were put in hand for night-time operations, including the *Lightning Bug* with 1.2 million candlepower searchlights mounted in the cabin doorway, the short-lived *Batship* with Low Light Level TV (LLLTV), and the more successful *Nighthawk* with starlight scope and searchlight.

Away from Vietnam, Quick Fix 1 was the title for ten Army EH-1H radio location and jamming helicopters converted from 1976. The US Navy's TH-1L trainer had the little-used name of Seawolf, and the 1968 Model 211 HueyTug was a one-off flying crane type converted from a UH-1C. Italian-built Agusta-Bell 205As in the Rhodesian/Zimbabwean Air Force were known as Cheetahs, and in the Japanese Ground Self-Defence Force the *Huey* has the local name Hiyodori (Bulbul) as well as the old American designation of HU-1.

The first twin-engined model was a UH-1D fitted with Continental turboshafts in 1965 to become the Model 208 Twin Delta, but it was soon overtaken by the Model 212 Twin Two-Twelve (military UH-1N) with a P&WC PT6T Twin Pac powerplant. Helitanker is the name used by Canadian company Conair for firefighting versions of both the single-engined Model 205 and 212 twin, with maximum water/retardant loads of 360

and 1,360 US gallons respectively. The addition of a four-blade rotor to the 212 produced the Model 412 in 1979, and from this Italian licensee Agusta developed the multi-role AB-412 Griffon; the same name was also chosen by Canada in 1993 for its CH-146 transport helicopter, a locally assembled 412.

A separate line of development from the original twin-blade/single-engine models began with the more powerful Model 214 (initially called the HueyPlus) of 1970; several were sold to Iran, but plans for large-scale production there as the Isfahan ended with the 1978 Islamic Revolution. A limited commercial offshoot was the Model 214B BigLifter, but more successful was the 214ST Super Transport (originally the ST denoted 'Stretched Twin') also once intended for Iranian production; with an 8ft stretch, up to twenty seats, and a new streamlined appearance, the line has come a long way from the original 8-'Grunt' (Infantryman) UH-1 of early Vietnam days. Perhaps even further removed is the American Aviation Corp. Penetrator, which gives the old *Huey* a new streamlined composite forward fuselage incorporating both gunship and transport features; prototype N3080W flew in 1991.

Hummer
Part of wider American slang and thought to derive from 'Humdinger', this is a popular term of appreciation for any aircraft that excels at its given task.

Hun
Although it evolved from North American's Sabre 45 project the F-100 Super Sabre was of course a totally different aircraft to the F-86, but pilots

everywhere knew it more familiarly anyway as the *Hun* or occasionally the *Silver Dollar*; nevertheless Danish pilots (at least) preserved a link by naming their F-100Ds *Super Dogs* in memory of the F-86 *Dog*. Surprisingly, the world's first supersonic fighter is sometimes referred to as being a **Lead Sled** although this is more associated with its time in Vietnam when it operated alongside faster types like the F-105 and F-4. Experienced pilots loved the F-100, but it was not a machine for the faint-hearted; at the best of times a landing (particularly in the flapless F-100A and C) was described as a 'controlled crash', and mishandling on the approach would often lead to the infamous 'Sabre Dance'. The least known of all Super Sabres must be the half-dozen RF-100A *Slick Chick* recce aircraft used on supersonic dashes over East German airspace.

Hunchback
Nickname of the Ilyushin I1-20, a 1948 *Shturmovik* type prototype, resembling a Short **Seamew** with its deep forward fuselage.

Hunchback Bomber
One of the more successful Italian types of WW2, the tri-motor bomber variants of the Savoia-Marchetti SM79 Sparviero (Sparrowhawk) rejoiced in the nickname of *Gobbo Maladetto* or *Hunchback*.

Hupmobile
US Navy nickname for the Piasecki HUP Retriever tandem-rotor helicopter (not to be confused with the Piasecki *Flying Banana*s). The US Army counterpart was the H-25 Army Mule.

Hurricane (i)
Provisional name, soon changed for obvious reasons, for the Bell XP-59A Airacomet, America's first jet aircraft. *Squirt* was probably inherited, along with the Frank Whittle/Power Jets technology which made the XP-59A possible, from Britain's E28/39 *Pioneer*.

Hurricane (ii)
Provisional name for the promising Hawker P1121, the F-4 Phantom-class fighter project cancelled in view of Government indifference in 1959.

Hurricane (iii)
To complement the Tornado, the name of **Hurricane** has been proposed in the UK for the controversial and much-delayed Anglo/German/ Spanish/Italian Eurofighter 2000, formerly the EFA or European Fighter Aircraft. The favoured name in Germany is reportedly Defender, whilst the leader in a straw poll by a British Sunday newspaper in 1993 was Eureka.

Hurricat
Well-known nickname for the 50 Sea Hurricane Mk IA fighters catapulted from specially converted merchant ships (CAM ships) for North Atlantic convoy protection from 1941, principally against marauding Fw 200 **Condor**s. After combat the pilot had the options of either bailing out or ditching alongside the convoy, or trying to reach the nearest land. Around 300 more Sea Hurricanes – or *Hooked Hurricanes* – were converted, principally by General Aircraft Ltd, and operated on mainly Russian convoys from the decks of converted merchant ships (MAC ships). The Hawker Hurricane is of course best remembered for its role in the Battle

of Britain, when it destroyed more enemy aircraft than all the other defences (including Spitfires) put together. A wartime variation on the more usual *Hurry* was *Hurryback*.

Husky
A mid-fifties update of surplus USAF Convair L-13 liaison aircraft, the Caribbean Traders Inc. **Husky** was offered with either the original Franklin powerplant, a Lycoming R-680 (**Husky** Mk II) or P&W R-975 (Mk III). The cabin of the **Husky** could accommodate eight seats, two stretcher-cases, or cargo, and the conversion sold in Latin American and the Caribbean area. A contemporary L-13 conversion was the six-seat Centaur with an enlarged vertical tail and choice of Lycoming or Jacobs engine, devised by Longren Aircraft of California.

Icarus
The Taylorcraft Aeroplanes (England) Plus D light aircraft was a slightly surprising choice as the new army Air Observation Post (AOP) type in 1941, and so it was perhaps in a state of shock that the fledgling company initially elected to call it the **Icarus**; not the most auspicious choice, in view of what happened to the fabled owner of that name. Fortunately, the name was quickly changed to one which was to become an inseparable part of British aviation for decades – Auster, from the Latin for 'South Wind'. After building roughly 1,600 AOP machines during the war the Taylorcraft company became disassociated with its former American partners and renamed itself Auster Aircraft in March 1946, at which point a new service name should strictly have been selected for succeeding military models.

Igor's Nightmare
Reference to the trials and tribulations of Igor Sikorsky in developing the Vought Sikorsky VS-300 helicopter, which finally made its first free flight in skeletal form at Stratford, Connecticut in May 1940. Among the problems encountered with the VS-300 was an early willingness to fly in any direction except forward; when questioned about it later, the Russian émigré replied with a twinkle in his eye that he nearly resorted to curing the problem by turning the pilot's seat around to face the other way. Though it was by no means the world's first helicopter to fly, the VS-300 was the first really practicable machine and the true forerunner of the helicopter as we know it today.

Imp
South African Air Force monicker for the much-loved Atlas Impala, a licence-built version of the Italian Aermacchi MB 326M jet trainer (Impala 1) or single-seat, ground-attack MB 326K (Impala 2). Embraer of Brazil also cut its teeth on jets by licence-building the EMB 326GB two-seat, trainer/ground-attack model, known to the Brazilian Air Force as the AT-26 and named Xavante (pronounced 'Shavante') after the Brazilian Indians.

Inca
Original name of the Piper PA-31 Navajo piston twin, and apparently still in use at the time of its maiden flight in September 1964. Although Piper fully exploited the basic design with their own stretched Chieftain (Navajo Chieftain prior to 1978), turboprop Cheyenne series, 'cross-kitted' Mojave, and unnamed T-1020 and T-1040 regional airliners, a number of independent conversions

have also found a ready market. Schafer Aircraft Modifications of Clifton, Texas specialise in fitting the ubiquitous P&WC PT6A turboprop, having begun in 1981 with a PT6A-135-powered Pressurised Navajo conversion known as the Comanchero. The Comanchero 500 is a Chieftain with either PT6A-20s (Comanchero 500A) or PT6A-27s (-500B), and in 1984 the conversion also became available in Brazil as the NE-821 Caraja (an indigenous Indian people) after Embraer transferred responsibility for their EMB-820C Navajo (licence-built Chieftain) to their Neiva subsidiary. Comanchero 750 is Schafer's title for a Cheyenne II conversion with 750shp PT6A-135s in place of the standard -28 units. Colemill Enterprises of Nashville offer a package including 350hp Textron Lycoming TIO-540 turbocharged piston engines, four-blade Hartzell props and optional 'Zip-Tip' winglets for both the Navajo and the Chieftain, the resulting conversions being named Panther and Panther II respectively.

Inexplicable
The Rohrbach-type metal construction of Beardmore's 157½ft-span research monoplane was often said to be more akin to shipbuilding practice than aviation, and ironically even its formal name of Inflexible sounds more like that of a Royal Navy dreadnought than an aeroplane. Already regarded as something of a white elephant before it flew at Martlesham Heath in early 1928, the Inflexible was lampooned mercilessly as the *Inexplicable*, *Impossible*, *Incredible* and *Brittle*.

Invisible Aeroplane
French term for the Rumpler **Taube** monoplane, after an aircraft flown by

Lt von Hiddeson bombed Paris in August 1914; a combination of bright sunlight and translucent fabric covering gave the German machine its unintentional 'stealth' characteristics.

Iron Bird
Not actually an aircraft at all, but the industry term for the ground rigs used to prove the flight control systems of modern airliners like the Airbus A340 and Boeing 777.

Iron Chicken
RAF nickname for the Westland Whirlwind helicopter, believed to have originated during the 1962-66 Indonesian confrontation when it was said to have 'clucked and flapped' its way around the jungle. The Whirlwind was of course a licence-built Sikorsky S-55 (the US Army's H-19 Chickasaw), a type also built by SNCASE of France with the slightly curious title of Elephant Joyeux.

Orlando Helicopters of Florida produce a host of imaginative OHA-S-55 conversions including the crop-spraying Bearcat, the Nite-Writer with a 40ft by 8ft electronic billboard, the passenger/ambulance Vistaplane with enlarged windows (and an optional one in the cabin floor) and the luxurious Heli-Camper with such accoutrements as a four-berth soundproofed cabin, shower, toilet, bar, and colour TV. Turbine-powered OHA-S-55T conversions, with the new powerplants relocated from the nose to the transmission deck behind the cockpit, are the Phoenix (P&WC PT6A) and Challenger (Textron Lycoming LTS101-700). In 1985, Orlando signed a twenty year deal for the OHA-S-55 range to be assembled at Guangzhou (Canton), China with the local name of **Panda**. Perhaps the most radical of all Orlando conversions is the US Army's QS-55 Aggressor or Hind Lookalike, acquired for training purposes and operable as either a drone or piloted aircraft. A new five-blade main rotor and composite-construction nose, tailboom and stub wings imitate the outline of the Soviet/CIS Mil Mi-24 Hind E, and there are even noise simulators to provide an authentic engine sound; the first of fifteen flew in the summer of 1988.

Iron Tadpole
Early nickname, along with the more enduring **Buff** for the US Navy's Grumman A-6 Intruder low-level strike aircraft and the EA-6B Prowler four-seat electronic warfare derivative. In the absence of a formal type name, the halfway-house EA-6A two-seater naturally became the *Protruder* (cf **Q-Bird**).

Jenny
One of the great names of American aviation, **Jenny** was simply a play on the Curtiss designation JN, but its ubiquity surely stems from the fact that it so perfectly matched the character of the biplane that became famous first as a WW1 trainer and later as a barnstorming aircraft. The Canadian-built JN-4Can evolved quite independently of the American JN-4, and was considered different enough to warrant the new and widely adhered-to title of Canuck.

Jet Dragon
Very early name, soon discarded, for the best-selling DH/HS/BAe 125 corporate jet, in tribute to its DH 84 and DH 89 biplane ancestors; likewise the RAF's navigation trainer variant was named Dominie like the military DH 89, but it should be noted that this does not apply to the RAF's transport

125s. The de Havilland pedigree was also exploited in the US, where the type was promoted as the DH 125 long after Hatfield had put the Hawker Siddeley title above the door. Later it was marketed by Beech as the Beech Hawker BH-125 and for some years the mark was generally known in North America simply as the Hawker; the name was formalised following the Raytheon takeover in 1993, when the current BAe 125-800 and BAe 1000 models became the Hawker 800 and Hawker 1000. Among foreign air force operators of the 125 are South Africa and Malaysia, with whom it is known as the Mercurius and Merpati respectively. A maritime variant of the HS 125-700 was offered in 1979 as the Protector with a MEL Marec radar in a new, unsightly, drooped nose radome; Protector 1 would have retained a limited transport capability whilst the Mk 2 was planned as a dedicated patrol variant, but neither were built.

Jet Spiteful
Very early name for what eventually became the Royal Navy's Supermarine Attacker jet fighter, using a basically similar laminar-flow wing to its piston-engined predecessor.

Jet Stork
Nickname of the Swiss FFA P-1604 (alias P-16) ground-attack fighter, tailored specifically for Alpine operations and first flown in August 1955. Production orders for the transonic *Dusenstorch* (*Jet Stork*) were later cancelled, but the basic wing layout lived on in the famous Learjet (cf **Transporter**).

Jet Stratoliner
Boeing's jet airliners are now so well-

known by their 7-7 designations that it is almost forgotten that the father of them all, the 707, originally had a name too (the original Stratoliner was the pre-war Model 307 with B-17 Flying Fortress wings). The Boeing Model number of 707 apparently had more appeal to the fare-paying public though, and in time the name atrophied, as did the self-explanatory sub-names of Intercontinental and Transcontinental. Nevertheless, several airlines chose to apply their own particular marketing names to the 707, such as AstroJet (American Airlines), Golden Jet (Continental), Jet Mainliner (United), Sunjet (Cyprus Airways), and perhaps most imaginatively Qantas, with their turbofan V-Jets, from 'vannus', the Latin for 'Fan'. The Comtran Super Q conversion centres around an engine hushkit installation to enable the 707 to meet FAR Part 36 noise rules, but also features an updated widebody-look cabin.

Military transport versions of the true 707 (the similar-looking KC-135 is dealt with under *Silver Sow*) include the USAF's unnamed C-137 series, and the Canadian Air Force's CC-137, which has the rarely used name of **Husky**. The 707-300 is also the basis for the E-3 Sentry early warning aircraft, occasionally nicknamed *Frisbee* because of the huge saucer-like rotodome, but perhaps most widely referred to by the acronym AWACS (Airborne Warning And Control System). Similarly, the name Phalcon is often used for Israel Aircraft Industries' AEW Boeing 707 conversion, but in fact the title properly belongs to its solid-state, phased-array radar system. Also based on the 707/E-3 airframe is the US Navy's E-6 TACAMO II (Take Charge And Move Out) used for Very

Low Frequency communications with submarines, via a trailed aerial no less than 4.9 miles long; the E-6 was originally named Hermes when the first was handed over to the Navy in 1988, but when this title failed to gain popular acceptance the Greek name was replaced by that of its Roman counterpart, Mercury. A further, but unnamed US Military variant is the E-8 (formerly EC-18C) carrying the JSTARS or Joint Stars (Joint Surveillance And Target Attack Reconnaissance System) which proved so valuable in the Gulf War.

Jet Trader II
Proposed commercial freight version of the McDonnell Douglas YC-15, two prototypes of which (along with a pair of Boeing YC-14s) were built in the mid-seventies for the abortive AMST (Advanced Medium STOL Transport) programme to replace the C-130 Hercules. The original Jet Trader was a DC-8 variant (cf *Diesel 8*).

Jug
Conventional wisdom has it that this universal nickname for the Republic P-47 Thunderbolt fighter was short for Juggernaut, while others claim it was named after the old tapered jugs used for moonshine liquor. In fact the name originated early in the P-47's service life as *Thunderjug*, which is American slang for a chamberpot; not the most auspicious of starts for what turned out to be a long and memorable line of Republic Thundercraft names, so could perhaps the juggernaut story have come from its Farmingdale, New York birthplace? British spotters knew the P-47 a little more prosaically as the *Flying Milkbottle* and the experimental 504 mph XP-47J was unofficially titled *Superman*.

Jumbo Jet

The British Press is sometimes credited with coining this slightly facetious name for the Boeing 747, supposedly around the time of the inaugural Pan Am New York–Heathrow service on 22 January 1970. In fact the name was in use by at least 1966 (the year the 747 was launched) and was initially taken as a term for the upcoming widebody generation as a whole, i.e. the 747, TriStar, DC-10 etc. Whatever the origin, there can be little doubt that the name is now a part of the English language, whilst the 747 itself has gone on to become one of the great symbols of the late twentieth century.

Junior Jumbo and *Baby Jumbo* were early nicknames for the long-range 747SP (special performance) with its fuselage shortened by over 47ft, whilst at the other end of the spectrum the 747-100SR (short range) demonstrator was at one stage billed as the 747 Super Airbus; the SR models were developed specifically for Japan Air Lines' ultra-short, island-hopping routes, with strengthened undercarriages to withstand more frequent take-offs and landings. Airline names have included Singapore Airlines' famous Big Tops (the 747-300 with the new stretched upper deck) and Mega Tops (747-400), Japan Air Lines' Sky Cruisers (747-400), Seaboard World's Containerships (747F freighters) and Qantas' Longreach, which not only reflects the 747-400's 8,400-mile range, but also commemorates the outback town where the airline began in 1920. Domestic military versions of the 747 are the E-4 Advanced Airborne National Command Post, and the Presidential VC-25A (cf *Air Force One*).

Jumbolino

Like most other commercial aircraft of its generation, Britain's successful short-haul jet has never had a proper name, just a succession of soulless designations like HS 146, BAe 146, and more recently the awful BAe RJ70, 80, 100 and 115, which refer to nothing more than the length of the tube and the number of seats squeezed into it. The nearest it came was when a marketing name was required for the Taiwan link-up, and what could have been more appropriate for an Eastern venture than **Dragon** or Superdragon (as suggested by *Aircraft Illustrated* magazine in 1991) which would also have recalled its famous Hatfield/de Havilland ancestor? The answer, apparently, was Avroliner. Fortunately there are still imaginative people in aviation such as Moritz Suter, founder of Swiss carrier Crossair, whose 146s bear the name **Jumbolino** along with a flying elephant motif; Australia's defunct Eastwest called its 146-300s Leisure Jets, whilst Air Wisconsin staff used the appropriate nickname of *Pudgie*. The nearest its makers have so far got to a formal title is the little-used Statesman name for the executive versions (the RAF's Queen's Flight BAe 146CC 2s are Statesmen in all but name) and the suffix of the 146QT overnight freighter stands for Quiet Trader.

Jump Jet

Along with **Hawk, Heron** and **Kestrel**, the now-famous Harrier name was first put forward in 1961 by Hawker's Senior Projects Engineer (and later aviation writer) Francis Mason for a possible RAF version of the P 1127 VTOL research aircraft. **Kestrel** was applied to the nine aircraft evaluated by the Tripartite Squadron (UK, West Germany and USA) at RAF West

Raynham from 1964, whilst Harrier was evidently earmarked for the larger, supersonic P 1154, then seen as the definitive VTOL fighter for both the RAF and RN; this fell victim to the spate of cancellations in the mid-sixties, and the RAF ended up after all with a heavily redesigned P 1127 (RAF), named Harrier in 1967.

Almost as soon as it made history by becoming the first fixed-wing VTOL type in service, with 1 Squadron RAF in 1969, it became known to Press and public alike as the *Jump Jet*, although in the RAF this was parodied as *Leaping Heap*, perhaps reflecting the fact that for all its supposed simplicity the first generation GR 1 and GR 3 was in many ways a complex little beast; *SNUF* (apparently inspired by the American *SLUF*) stands for 'Smelly, Noisy, Ugly ... Feller'? *Shar* is simply shorthand for the RN's Sea Harrier (known in the embryonic stage as the Maritime Harrier) which during the 1982 Falklands War was gleefully dubbed *Black Death* by the British tabloid Press on behalf of the Argentine Air Force. Matador is the Spanish Navy name for both the first generation AV-8S and the current EAV-8B Harrier II. Other foreign customers include India and Italy, but perhaps the biggest coup of all remains the original sale to America in 1969, the significance of which was neatly illustrated by a notice which reputedly appeared at NAS Cherry Point: 'The Harrier was invented by the Marine Corps and sold to the British; any statement to the contrary is pure twaddle'. American nicknames include *Whistling Shitcan* and *Scarier*.

Justicialista del Aire
Former semi-official name of the Argentinian FMA/Dinfia IA 35 Huanquero light transport/crew

trainer, changed following the overthrow of Peron's Justicialista Party in 1955. A ten-seat civil prototype flew in 1960 as the Pandora, and another Huanquero was experimentally fitted with Turbomeca Bastan turboprops to become the Constancia II.

Kangaroo (i)
A distant relative of the RAF's spectacular Fairey Fox bomber, the Fox VII was a single-seat fighter biplane built by Belgian subsidiary Avions Fairey, and earned its nickname because of its large underslung radiator 'pouch'. Two were built in 1935, and as a single-place version of the production Fox VI two-seater the type was also known as the Mono-Fox.

Kangaroo (ii)
Predictable but wholly unofficial name sometimes applied to the Commonwealth CA-15 fighter prototype which, looking like a Mustang on steroids, was flown in Australia for the first time in March 1946; equally unofficial was an allegedly recorded level speed of no less than 502.2 mph.

Kangaroo (iii)
Informal name, evidently taken from the Winnie the Pooh books of A.A. Milne, for the Miles M68 Boxcar, an innovative four-engined light freighter with a detachable cargo pod. A single prototype (G-AJJM) was flown, but the project ended with the collapse of the Miles company in the bitter winter of 1947.

Kania
The Soviet-designed Mil Mi-2 (NATO – Hoplite) light turbine helicopter was manufactured exclusively in Poland by

PZL Swidnik, who also developed a number of new versions, including the Bazant (Pheasant) agricultural model with twin external hoppers carrying up to 500 litres of chemicals or fertiliser. More ambitious was the westernised **Kania** (**Kite**) of 1979, with twin Allison 250-C20B engines. This was initially marketed in the West as the Kitty Hawk, but a version with more powerful Allisons was also offered as the Taurus by Spitfire Helicopters of Pennsylvania; the same company also offered the basic Mi-2 as the Taurus II. Certification problems plagued the Westernised versions however, and the whole Mi-2/Kania programme was temporarily suspended in the early nineties after the cessation of Mi-2 orders from the former Soviet Union.

Katiuska
Name used by Republican forces in the Spanish Civil War for the Soviet-supplied Tupolev SB-2 twin-engined light bombers, after a Russian girl in a contemporary light opera. Reports at the time often referred to the SB-2 as the 'Martin' in the widely-held belief that the Republicans were operating American Martin 167s. After the civil war, the SB-2 was taken on by the Spanish Air Force as the *Sofia*.

Kestrel
Name of the advanced trainer designed as a private venture by Miles (Phillips & Powis) and belatedly ordered for the RAF in June 1938 as the Master.

Kfir
The saga of the Israeli Mirage III/5 developments began with de Gaulle's embargo of the 50 ordered and paid for Mirage 5J (for Jewish) fighters, as a response to the 1967 Six Day War.

Already a Mirage III operator, Israel began by studying a challenging development with a General Electric J79 engine (as used in the F-4 Phantom) but in the interim devised their own pirated copy of the original Mirage 5. With the alleged complicity of Mossad (the Israeli intelligence agency) plans were obtained through industrial espionage (including the famous Sulzer case in Switzerland involving the Atar 9C engine), and the first Israel Aircraft Industries Nesher (Eagle) flew in September 1969; differences from the French original were minimal, but included a Martin Baker ejection seat and Sidewinder or Shafrir (Dragonfly) missiles. Approximately 100 Neshers were delivered to the Israeli Air Force from 1971, and 39 refurbished aircraft were sold to Argentina from 1978 as Daggers; survivors of the Falklands War were updated with new Israeli avionics in 1984 to produce a more closely optimised ground-attack aircraft, and given the somewhat curious title of Finger.

Meanwhile, on 21 September 1970 the legendary Israeli test pilot Danny Shapira had flown a purely experimental two-seat Mirage IIIBJ adapted to take a J79 engine and renamed Technolog, but leaked reports of a similarly powered variant supposedly named Barak (Lightning) taking part in the 1973 Yom Kippur War turned out to have been a deception; Barak never existed and likewise a straightforward J79-powered Nesher known as the Ra'am (Thunderbolt) progressed no further than a single prototype flown in June 1973. Instead, effort was concentrated on an ambitious project marrying the J79 to a more advanced airframe development with a shortened but wider rear fuselage, intake/spine at the base of the fin, modified undercarriage, and upgraded weapons system; initially named Rackdan (Dancer) the new and definitive multi-role fighter entered service in 1975 as the **Kfir** (Lion Cub). Exports have been hampered by political constraints, but among foreign operators were the US Navy and Marine Corps who between 1985 and 1989 leased a total of 25 Kfir C-1s (lacking the full-size canards which later became a hallmark of the type) as interim aggressor aircraft; these were given the US designation of F-21A and the informal but more pronounceable name of *Lion*.

Nammer (Tiger) is the name of a proposed Kfir/Mirage III/5 update by Israel Aircraft Industries in the nineties, with a new GE F404 engine and modern weapons system (cf **Milan** and **Balzac**).

Killer
Untold numbers of aircraft have borne this epithet at some time in their careers, but the Soviet Yak-3U was unusual in that *Ubiytsa* (*Killer*) referred not to any reputation for accidents but to its prowess as a fighter. Very much pilots' aeroplanes, the wartime Yak fighters with their inline engines were also given the name *Ostronosyi* (*Sharp Nose*) by their crews to set them apart from larger radial-engined types like the Lavochkin La-5. The final wartime development, the Yak-9 (NATO – Frank) also bore the officially-sanctioned nickname of *Yastrebok* or *Little Hawk*.

King Arthur
In contrast to other BEA class names, **King Arthur** referred not to a specific type of aircraft but to its helicopter fleet as a whole. The title was

introduced along with a clutch of other BEA class names in 1950, and was discontinued around a decade later. Individual helicopters within BEAH were named after Arthurian knights, and the types involved were the Sikorsky S-51, the Bell 47B, G, and J, the Bristol 171 Sycamore and 173, and the Westland-Sikorsky S-55 Whirlwind.

King Cat
One of a multitude of engine-change conversions applied to the Grumman G-164 Ag-Cat agricultural biplane (acquired by Schweizer in 1981), the Mid-Continent King Cat package is based around the installation of a 1,200hp Wright Cyclone R-1820 radial in the G-164A Super Ag-Cat. Another attempt to beat a growing shortage of the Super Ag-Cat's standard 600hp P&W R-1340 engine was the 1980 Twin Cat conversion devised by the aptly named Twin Cat Corp, with a new solid nosecone and a pair of 310hp Lycoming flat-sixes attached to the lower wings. Ethiopian Airlines produce the Ag-Cat Super B Turbine (PT6A turboprop) under licence as the Eshet, the first of which was rolled out in December 1986.

Kingston
Original name, after its Kingston-upon-Thames birthplace, for the RAF's twenties-vintage Hawker Horsley day bomber; the final name came from that of Tommy Sopwith's former home, Horsley Towers. Two torpedo bomber variants were built for Denmark in 1932 with the name Dantorp, but plans for further licence-manufacture in Copenhagen failed to materialise.

Kiowa
Bell's OH-4A (Model 206) lost the

original US Army LOH contract to Hughes (cf *Loach*) in 1965, but from it they derived the popular Model 206A JetRanger, arguably the most familiar of today's civil helicopters. In any case, the LOH competition was re-opened in 1967 and won by a militarised JetRanger which was ordered into large-scale production as the OH-58 **Kiowa**. Another early military customer was the US Navy who in 1968 took delivery of an initial batch of 40 as TH-57 SeaRanger trainers; in contrast, the Army did not adopt a trainer version until 1993, when a variant of the commercial Model 206B-3 was selected as the TH-67 Creek.

A major modification to the Army's OH-58 **Kiowa** inventory under AHIP (the Army Helicopter Improvement Programme) materialised in 1983 as the OH-58D Advanced Scout (initially the Aeroscout) with an uprated engine, new four-blade rotor (hence the new Bell Model number, 406) and a sophisticated sensor package in the Mast Mounted Sight (MMS) 'football' atop the rotor hub. The Model 406CS Combat Scout is essentially a simplified export version of the OH-58D, lacking the MMS and some of the more sophisticated avionics; first deliveries were to the Royal Saudi Land Forces in 1990. The OH-58D was initially intended for general armed scout duties, but a more heavily-armed attack version (Hellfire and Stinger missiles) was quickly developed in 1987 and deployed by Special Forces in the Persian Gulf to guard international shipping against nocturnal attacks by Iranian fast patrol boats; such was the success of these fifteen *Night Owls*, as they were unofficially known, that it was decided to further update most of the OH-58D fleet to fully-armed configuration, in

which form they were renamed Kiowa Warrior (in practice usually just **Warrior**) in 1990.

Amongst the numerous foreign operators of the basic military **Kiowa** was Australia, where in 1972 the Army Minister, Robert Katter, took it upon himself to rename it Kalkadoon after a Queensland Aboriginal group; disgruntled Army fliers responded by christening their aircraft *Kattercopters* in his honour, and following a change of Government the American title of **Kiowa** was quietly reinstated. Other Aussie nicknames included *Sky Scooter*, and the disparaging *Pop-up Target*.

JetRanger conversions over the years have included the Imagineering Systems (Texas) Pathfinder 206 trainer with revised seating, crashworthy fuel system, and a new tricycle-wheeled undercarriage in place of the traditional skids; a prototype was completed in 1989 but failed to attract the Army. Also in 1989, Iran unveiled the Zafar 300, a dedicated two-seat tandem gunship derived from the Model 206 by Seyedo Shohada, but the exact status of the project remains uncertain.

A mainstream development of the JetRanger was the stretched seven-seat Model 206L LongRanger successfully launched by Bell in 1973, but the military TexasRanger version developed as a private venture from 1980 failed to find a buyer despite an extensive marketing effort. Global Helicopter Technology of Fort Worth also offer an armed LongRanger conversion as the NightRanger, with Forward-Looking Infra-Red (FLIR), weapons pylons from the Model 406CS Combat Scout, and pintle-mounted cabin machine-guns. However, the most successful LongRanger conversion to date is the

civilian Gemini ST converted by
Tridair of California to take a new
twin-Allison turbine powerplant in
place of the usual single engine; the
first flew in 1990 and an initial batch
of 50 conversions was planned. Bell
themselves took out a licence to use
the installation in new-build
LongRangers, in which form it is
known as the Model 206LT
TwinRanger (not to be confused with
the wholly new Model 400
TwinRanger of 1984, development of
which was suspended on commercial
grounds).

Kipper Kite
Traditional RAF slang for maritime
patrol aircraft, dating back at least as
far as WW2.

Kite
The more romantic and the more
fanciful may have looked to the birds
for inspiration, but of course the
simple kite (and not the ornithopter)
was Man's first heavier-than-air flying
machine; as such it is perhaps only
fitting that **Kite** apparently predates
even *Bird* as a really popular slang
term for aircraft in general. Although
Hargrave invented the boxkite as long
ago as 1893, it was not until the
appearance of the much-cloned Henri
Farman biplanes of 1909 (the Bristol
Boxkite was but one copy) that that
variation really became established;
ironically so, as the Henri Farmans
were actually instrumental in the
move away from the more box-like
structures with 'side-curtains' between
the flying surfaces.

The slightly sardonic terms *Bus*,
Ship, and *Crate* all date basically from
WW1 and speak for themselves, but in
the French Aviation Militaire they
spoke of aeroplanes as *Zincs* (the
colloquial term for French cafés with

galvanised tabletops), whilst Kiwis in
the RFC were sometimes heard to talk
of *Grids*, as exemplified by Major
Keith 'Grid' Caldwell of 60 and 74
Squadrons.

Knuckleduster
The Short R24/31 flying boat acquired
its aggressive nickname because the
position of the twin Rolls-Royce
Goshawk VIII engines at the joint or
'knuckle' of the gull wing gave it an
appearance like the weapon of that
name. Designated S18 by Shorts, the
single example (K3574) flew for the
first time in 1934.

Komadori
T-3 **Komadori** (Robin) is the Japanese
Air Self-Defence Force (JASDF) title
for the Beech T-34 Mentor trainer,
built under licence by Fuji. The same
company also developed the
'widebody' LM-1 and LM-2 Nikko
four-seat liaison variant, which is
known to the Ground SDF as the
Harukaze (Spring Breeze).

Kong Yun
Kong Yun (Cloud) was a French-
assisted upgrade to China's Qiang-5
(NATO – Fantan), the long-running
ground-attack derivative of the MiG-
19 which differs from the original
Soviet fighter in its long, pointed nose
and side-mounted intakes.
Incorporating an extensive avionics
update and bearing the Western-style
designation A-5K (Qiang-5 means
'attack aircraft No. 5') the first of two
prototypes flew in September 1988 but
the programme was discontinued two
years later. A parallel programme
aimed more at the export market was
the Italian-assisted A-5M, with an
avionics package based on that of the
Italo/Brazilian AMX.

In Vietnam, the original MiG-19

fighter (NATO – Farmer) was given
the identification code White Bandit
by the US.

Kukko
Finnish Air Force name for the fifteen
Gloster Gamecock II fighter biplanes
built under licence by the National
Aircraft Factory at Helsinki from
1929. The RAF's Gamecock I
inherited the wing-flutter problems of
its predecessor the Grebe, and a
partial cure was effected by fitting
additional outboard 'vee-struts'; in
this form it was sometimes referred to
as the *Cat's Cradle* after the children's
game involving string stretched in
complex patterns over the fingers.

Kurier
Churchill branded it the '*Scourge of
the Atlantic*', and *Tinfoil Bomber* was
a British jibe at its airliner origins and
early structural problems, but the
mystery of the Focke-Wulf Fw 200
Condor is why the British aviation
Press (*Aeroplane*, *Spotter*, even *Jane's*
as late as 1943/44) persisted in
claiming the maritime bomber
variants were known as the Fw 200K
Kurier. To paper over a lack of
information (some would say a lack of
effort) it was common practice to tack
a spurious K onto commercially-
derived Luftwaffe types, but this was
usually intended to suggest a Krieg or
'War' version; and if they had to
invent a name, surely **Kurier** would
have better suited the pre-war airliner
model?

Lama
Built under licence in India as the
HAL Cheetah (not to be confused
with the **Chetak**) and by Helibras in
Brazil as the Gaviao (**Kestrel**) the
Aerospatiale SA 315B **Lama** of 1969
was a 'hot and high' development of

the established Alouette (Lark) II
helicopter, utilising the transmission
system of the larger Alouette III. The
1957 SE 3131 Governeur represented
an attempt to introduce a four-
passenger executive model of the
Alouette II, in which the familiar open
structure was replaced by new
streamlined bodywork designed by
automobile stylist Raymond Loewy.
In Algeria in 1958 the Alouette II
became the first turbine helicopter to
go into combat, armed with machine-
guns and anti-tank missiles, and
earned the nickname the *Horned
Butterfly*.

Lanc
Synonymous with what is arguably the
most famous bomber aircraft in
history, **Lanc** requires no comment,
although perhaps forgotten now is the
wartime variation *Lanky*. The
Lancaster was of course born as a
result of the disappointing twin-
engined Avro Manchester, and was
very briefly known as the Manchester
III around the time of its maiden flight
on 9 January 1941. The downside of
the Lancaster's deserved fame was
that the wartime media tended to
ignore the very real contribution made
by other RAF bombers, which led to
the cynical title of *Daily Mirror
Aircraft* among Halifax crews. The
first civilianised Lancaster was Trans
Canada Airlines' CF-CMS as early as
1943, and from 1945 Avro produced
over 80 Type 691 Lancastrians with
faired-over noses and tails, and
sideways seating for either nine or
thirteen. Strictly an interim post-war
transport, the Lancastrian served with
the RAF and such airlines as BOAC,
British South American Airways
Corp. (BSAAC), Silver City, Skyways
and Alitalia.

Lancier

Also referred to as the Alpha Jet 3, the 1987 **Lancier** was to have been an optimised attack version of the Dassault/Dornier jet trainer with Thomson-CSF Agave multi-mode radar and a state-of-the-art cockpit; no orders were forthcoming, and the **Lancier** was cancelled.

Lead Sled

Widely used American term for any aircraft considered to be underpowered and unresponsive, lacking in 'oomph'.

Learstar

Just as the commercial Lockheed Model 14 Super Electra sired the RAF's Hudson (cf *Boomerang*), so the stretched Model 18 Lodestar formed the basis of such well-documented military types as the US Navy's PV-1 Ventura (the *Pregnant Pig* in the RAF), the USAAF's B-34 and B-37 Lexington, and the re-winged PV-2 **Harpoon**. Among numerous post-war executive conversions exploiting the speed and capacity of the series were the Howard Super Ventura and Models 250, 350 and 500, differing in power, wing-span and fuel capacity; the top of the line Howard 500 first flew in 1959 and was certificated in February 1963 as a new aircraft, rather than being given the Supplemental Type Certificate (STC) usually applicable to conversions. Lear/ Pacaero's 1954 **Learstar** offered a cruising speed of around 280 mph, a luxurious twelve-seat interior and extensively updated systems, and Pacific Airmotive applied a broadly similar concept to the PV-2 **Harpoon** airframe to produce their eight- to fourteen-seater pressurised Centaurus. The *Ventellation* was another in the series of Lockheed

hybrids, namely the fifth production Ventura retained by the company to test-fly the engine installation of the Constellation.

Lightning II

Whilst **Lightning II** (after the wartime P-38) is clear favourite to become the official name of the Lockheed F-22 Advanced Tactical Fighter (ATF) a certain General in the Pentagon has campaigned to continue the Lockheed tradition of stellar titles by calling it SuperStar; as Uncle Roger commented in 'Budgie News' (*Flight International*) fans of West End musicals can predict the response if it is formally adopted . . .

Limousine

Sometimes referred to as the *Coffee Pot* on account of its prominent vertical exhaust stack, the Breguet 14 bomber, like its British counterpart the DH 4, was the subject of several civilian conversions after WW1. The CMA group devised the Breguet 14T2 Salon with a deepened forward fuselage housing a two-seat cabin, large rectangular side windows, repositioned cockpit, and fuel in pods between the wings. The enlarged four-passenger Breguet 18T Berline with three-bay wings ran to only two confirmed examples, but some features of it found their way onto the 14Tbis **Limousine** of 1921, which was easily distinguishable by the row of four circular windows each side of the upper cabin. Lignes Aeriennes Latecoere ('The Line') operated at least 190 Breguet 14s at various times, many on mail routes to Africa and South America (as immortalised by the novels of Saint-Exupéry) and also developed their own conversion with a single open cockpit and underwing panniers, known as the Breguet

Latecoere or (perhaps a little misleadingly) as the Torpedo (Open Tourer).

Lincolnian
Less well-known than either the Halton or Lancastrian conversions after WW2 was the Avro 695 Lincolnian freighter based on the Lincoln bomber. Just five were produced; one as an Antarctic survey aircraft for Argentina; another (converted by Airflight Ltd) was used by Surrey Flying Services on the Berlin airlift; and three with deepened fuselages were scrapped when their intended role as meat transports in Paraguay failed to develop. Named after 'Bomber County's' principal city, the Avro 694 Lincoln bomber was a development of the famous Lancaster, and was briefly known in the planning stage as the Lancaster Mk IV (Rolls-Royce Merlin 85s) or Mk V (Packard Merlins).

Linnet
Well-known anglicised version of the French Piel Emeraude two-seat, low wing lightplane; five **Linnet**s only were completed, two (G-APNS and APRH) at White Waltham by Garland Bianchi, and three (G-ASFW, ASMT and ASZR) by Fairtravel Ltd between 1963 and 1965. The attractive Emeraude was also built in South Africa by General Aircraft (later Southern Aircraft & Robertson Aircraft) as the Aeriel 2, and in West Germany by Binder Aviatik as the Smaragd (Emerald) with a sliding canopy and enlarged fin. Piel's CP-60 series Diamant was a three-seater development for amateur construction.

Little Annie
The Antonov An-2 utility transport

(NATO – Colt) may be unique as the only biplane designed and built in quantity since the war, but in many ways it is also typically Russian – old-fashioned, functional, and seemingly immortal. Around the time of its maiden flight in 1947 the An-2, built to a specification issued by the Agriculture Ministry, was known as the Kolkhoznik (Collective Farmer), but this name appears to have been discarded at an early stage; the version built in China from 1957 was named Fongshu 2 (Harvester 2). In its homeland the An-2 also has the affectionate title of *Annushka* (*Little Annie*), whilst in the West it was dubbed *Big Ant* in recognition of both its design bureau and its sheer bulk for a biplane. The same term was also applied in more recent years to the vastly different (in all senses) An-124 Ruslan (NATO – Condor).

After roughly 5,000 had been built in the Soviet Union, the major production effort was switched to Poland in 1960, where approximately 10,000 more were built up to the mid-eighties. Polish variants included the Geofiz geophysical survey version, and the remarkable Lala-1 research aircraft, the name of which derived from Latajace Laboratorium, or Flying Laboratory. Built as a test-bed for the M-15 Belphagor agricultural aircraft (the world's only jet-propelled biplane, and likely to remain so) the Lala-1 retained the usual radial engine in the nose, but also had an Ivchenko AI-25 turbofan in a truncated rear fuselage, fed by a single starboard-mounted intake; a reverse-tricycle undercarriage was fitted, and a new twin-fin tail unit was carried clear of the jet efflux by upswept outriggers.

Little Falcon
Nickname of the 1942 Fiat C6B

biplane trainer, resembling a scaled-down CR42 Falco fighter.

Little Shaver

Although quickly abandoned by the British, who originally called it the Caribou, the American-built Bell P-39 Airacobra (noted for its mid-mounted engine) was considered superior even to the Spitfire in the Soviet Union, where its pilots affectionately dubbed it *Britchik or Little Shaver* (shaving being slang for low-level strafing). The XFL-1 Airabonita was a navalised variant with taildragger undercarriage and redesigned tail, abandoned after a single prototype had been flown in 1940.

Lizzie (i)

They used to say that there were only two types of aeroplane – Lysanders and The Rest—and certainly there can have been few better known early war types than Westland's army cooperation monoplane with its high double-tapered wing, tapered fuselage (hence *Flying Carrot*) and huge spatted undercarriage. Stories about it are legion, but it was in cloak and dagger Mk III form that it was truly immortalised, landing SOE Agents ('Joes') into occupied territory under cover of darkness. Amongst numerous experimental versions were the anti-invasion *Pregnant Perch* (L4673) with a ventral gun position in a bulged rear fuselage, and the P12 Tandem Wing Lysander or Wendover (K6127) inspired by the French Delanne designs, which featured a second smaller rear wing carrying endplate fins, and a gun turret (only a mock-up turret was flown) on the end of the truncated rear fuselage.

Lizzie (ii)

All twenty production Airspeed AS57

Ambassadors were delivered new to BEA, and consequently the twin-engined airliner became generally better known by the airline's class name of Elizabethan; the manufacturer's type name only really came into its own when BEA began disposing its fleet to other operators. A plan to re-equip part of the BEA fleet with Napier Eland turboprops would have resulted in the name of New Elizabethan, reflecting not only the modification but also the accession of Queen Elizabeth II in 1952; in the event only one aircraft (G-ALFR 'Golden Hind') was converted. Graceful, and popular with passengers, the *Lizzie* will sadly be remembered by many as the type (G-ALZU 'Lord Burghley') involved in the Manchester United 'Busby Babes' crash at Munich on 6 February 1958. A military freighter variant was unveiled in model form at the 1946 SBAC show as the AS60 Ayrshire, but was not built.

Loach

The Hughes (now McDonnell Douglas Helicopters) Model 369M was a surprising but worthy winner of the original US Army Light Observation Helicopter (LOH) competition in 1965, and it was from those initials that the resultant OH-6 Cayuse took its most popular nickname; *Sperm* ran it a close second in Vietnam, and along with *Egg* succinctly describes the look of the compact OH-6. In commercial and military export form the type was known as the Model 500, and since the improved 500D of 1974 with a new five-blade rotor and T-tail, subsequent military developments have borne the name Defender. A single OH-6A was revealed in 1971 to have been modified with a five-blade main rotor with re-profiled tips, four-blade tail rotor, and sound blanketing around the entire powerplant assembly, making The Quiet One, as it was formally known, an early essay in helicopter stealth technology. The 30-plus MH-6Js used for Special Operations by the US Army's Task Force 160 'Night Stalkers' are derivatives of the MD 530MG Defender, and are sometimes referred to by the name Nightfox; the MH-6J designation superseded the former AH-6G (gunship) and MH-6H (covert transport) designations used prior to modification of the fleet to a common standard. The civilian MD 530F Lifter is a 'hot and high' derivative of the Model 500E series which in 1982 first introduced the new longer and more streamlined nose profile.

Lone Ranger
Due purely to changing requirements, just a single Boeing XPBB-1 Sea Ranger flying boat was built, flying for the first time in July 1942. With a span of 139ft 8½in the Sea Ranger was the largest twin-engined aircraft of its day, and endurance was estimated at no less than 72 hours.

Flying Longhouse
Local name for the RAF's Westland (ex-Bristol) Belvedere tandem-rotor helicopter during the Indonesian confrontation in the sixties. One of the Belvedere's forerunners, the Bristol 173 Mk 2 (G-AMJI) intended for BEA and fitted with fore-and-aft wings to offload the rotors, was briefly known as the Rotorcoach, an appropriate title for what someone once likened (in wingless form) to a piece of Bakerloo tube train having become airborne.

Loose, Wobbly & Frail
Colloquial term for the Model V trainer/observation biplane built at College Point, NY in WW1 by the L.W.F. Engineering Co. (the initials actually stood for Lowe, Willard and Fowler).

Flying LST
Obvious nickname (an LST being a Landing Ship, Tank) for the intriguing Convair R3Y-2 Tradewind, a four-turboprop flying boat with nose-loading doors to allow heavy loads to be landed in a beach-head assault. The 145ft-span Tradewind was originally designed as a long-range patrol/anti-submarine 'boat, in which form a single XP5Y-1 flew for the first time in 1950; changing requirements limited further production for the US Navy to six R3Y-2s and five R3Y-1 transports without the bow-doors.

Lucy
American nickname for the Caudron C272 Luciole, a two-seat French training biplane built in quantity in the thirties.

Luna Commander
Name originally proposed for the classy Rockwell Commander 112 single-engined light aircraft, presumably to cash in on the company's extensive involvement in the contemporary Apollo space programme. Shortly before Rockwell suspended production of its single-engined line in 1979, the Model 112TC-A and more powerful Model 114 were briefly named Alpine Commander and Gran Turismo respectively.

Maggie
As an RAF war-time trainer, the Miles M14 Magister came second only

to the immortal Tiger Moth in terms of the affection in which it was held. The Magister primary trainer was a fully-militarised development of the pre-war M2 Hawk monoplane (originally known as the Ibex, but changed to avoid confusion with Bert Hinkler's Ibis), and after the war numerous *Maggie*s were demobbed with the title Hawk Trainer Mk III. *Maggiebomber* was the title used for about fifteen equipped with bomb racks in 1940 to harass ground forces in the event of invasion.

Mainliner
From the mid-thirties until a re-vamp in the seventies, America's huge United Air Lines (UAL) promoted itself as the Main Line and its aircraft as Mainliners. Among the types carrying the title, with varying degrees of prominence, were the Boeing 247, Douglas DC-3, DC-4, DC-6 and DC-7, Boeing 377 Stratocruiser, and Convair CV-240 propliners. A variation was the Super Mainliner title adorning the original triple-tailed Douglas DC-4 (NC18100) of 1938, abandoned because of its complexity; this was retroactively designated DC-4E (for Experimental) and is not to be confused with the later DC-4 and military C-54 series. Jet Mainliners included the Sud Caravelle and the Boeing 707 and 720.

Mango
Resembling a cross between a Messerschmitt Bubble Car and a Piper **Cherokee**, with pram parts thrown in for good measure, the camouflaged Transavia PL12 M300 was publicly unveiled at the 1988 Australian Bicentennial Air Show as a light military derivative of the long-established Airtruk/Skyfarmer/Bushranger line.

Mariner
A 1943 proposal for a maritime patrol version of the Miles M38 Messenger light aircraft, the M38A **Mariner** would have been operated from small decks built onto merchant ships as a counter to the U-boat threat.

Mars
Generic name for the family of biplanes derived mainly from the late WW1-vintage Nighthawk fighter by the fledging Gloster company, which acquired the rights and a large component inventory from Cricklewood-based Nieuport & General when they closed in 1920. The original short-span Mars I racer (G-EAXZ) was better known as the *Bamel* after a remark by designer Henry Folland during construction that with its partly uncovered fuselage and its fuel tank 'hump' it looked 'half bare, half Camel'; later it was rebuilt as the Gloster I. Mars II, III and IV were alternative names for the 50 Sparrow-hawk I, II and III fighters built for the Japanese Navy, but there was also a single Mars III (G-EAYN) retained as a company demonstrator/racer. The Mars V two-seater remained a paper project, as did the single-seat Mars VII; the Mars VI Nighthawk was an export re-hash of the WW1 version which sold to Greece (25), and 22 Mars X Nightjars originally intended as naval fighters served with the RAF between 1922 and 1924. Mars VIII and Mars IX were unbuilt transport projects, unrelated to the Nighthawk fighter and its Gloster derivatives. A separate racer development of the Nighthawk by original designers Nieuport & General ('British Nieuport') was the one-off Nieuhawk (G-EAJY) of 1919.

Marsupiale
Widely used term, along with *Canguro*

(*Kangaroo*) for both the Savoia-Marchetti SM-75 and SM-82, Italian tri-motor transports noted for their unusually deep and capacious fuselages.

Martinet
Having built the German Siebel Si 204 twin-engined transport during the occupation of France, SNCAC and SNCAN went on to build a further 350 up to 1949 as the NC 701 and NC 702 **Martinet (Swift)**.

Martlet (i)
Provisional name for the attractive Beagle M218 light twin, first flown in August 1962 with the Class B registration G-35-6 (later G-ASCK). The name clearly reflected its Miles origins, but had the M218 gone into production it would most likely have been changed to one more in keeping with the Beagle 'Kennel Club'.

Martlet (ii)
Strizh (also translated as *Swift*) is the semi-official name appearing latterly in the Russian Air Force for the Sukhoi Su-17 swing-wing strike aircraft (NATO – Fitter C, D, E, G, H, K). The similar Su-22M-4 (NATO – Fitter K) was known in the former East German Air Force as the *Humpback*.

Maryland
RAF-only name for the Martin 167W twin-engined light bomber, rejected by the USAAF after evaluation as the XA-22, but also supplied in small numbers to France.

Maserati
In the sixties the Convair jet airliners had the dubious distinction of two separate entries in the Guinness Book of Records, one for being the world's

fastest airliners, and the other for the crippling losses they caused Convair; unable to compete effectively with the larger 707/720 and DC-8, just 65 Model 880s and 37 Model 990s were built, and the General Dynamics subsidiary lost something like $425 million on the programme. Originally known as the Model 22 Skylark, the smaller of the pair was later named Golden Arrow in connection with a gimmicky scheme to give it a gold-tinted exterior skin; this in turn was dropped, and the type finally flew in January 1959 as the Model 880, a designation based on the cruising speed in feet per second. The Forty-Niner was an unbuilt short-haul development, circa 1960, and in the late seventies American Jet Industries (AJI) used the name Airlifter for its cargo conversion with port-side cargo door, reinforced floor, and an impressive 51,000lb payload; eleven were converted but only four sold. The stretched longer-range Model 990, instantly recognisable by its overwing anti-shock bodies ('Kuchemann Carrots') was given the famous name of Coronado by the Swissair/SAS consortium. Powered by General Electric CJ805s, a commercial version of the notoriously 'dirty' J79 (as used in the F-4, F-104, etc.) the Convair jet airliners were known to their crews as *Smoking Joes*, and of course *Maserati* was an allusion to the breed of racing car.

Flying Mattress
Nothing to do with the contemporary *Flying Bedstead*, this was a popular nickname for the Cardington Kite, a low-powered, inflatable-wing aircraft intended to provide the British Army with a cheap communication and observation machine.

Meatbox

One of the more unusual aspects concerning the development of the Gloster **Meteor**, Britain's and the Allies' first operational jet fighter, was the trouble involved in finding a name for it. In early 1941 the Ministry of Aircraft Production (MAP) suggested Cyclone, Scourge, Tempest, Terrific, Terrifier, Terrifire, or Thunderbolt for what was then being referred to as the F9/40 Gloster Whittle aeroplane, and when these proposals failed to impress Gloster the names Avenger, Dauntless, Skyrocket, Tyrant, Violent, Vortex, Wildfire and Wrathful were all bandied back and forth between the two. Gloster apparently preferred Avenger, but the MAP ruled it out on the grounds that it would clash with the American Vultee Vengeance and announced instead that Thunderbolt had been officially selected, to which Gloster duly replied by pointing out the Republic P-47 Thunderbolt to the men in bowler hats. At some stage in the proceedings Gloster also apparently favoured the name Rampage, to the extent that the third prototype (DG204/G) with underslung Metrovick F2 engines was evidently designated F9/40M Rampage 2 for a while. The MAP finally announced the choice of the name **Meteor** in February 1942 and this time stuck to their guns, despite a last-ditch attempt by Gloster to have it changed to Ace, Annihilator or Reaper.

Despite (or maybe even because of) the wrangling over the name, the irreverent pun *Meatbox* became popular post-war, although many pilots were, and remain, openly hostile to the name. A more sardonic title was *Phantom Diver*, earned by the Meteor T7 trainer after a string of mysterious crashes in the mid-fifties.

The cause was found to be a particular combination (what would be known in the US as a 'Coffin Corner' of the flight envelope) of turning onto the downwind leg for landing with the dive-brakes open and then selecting undercarriage down, resulting in a stall; the problem was peculiar to the T7 version, and returned with a vengeance as late as May 1988, when it caused the crash of a preserved T7 during a display at Coventry airport.

Star Meteor was the title of a pair of F4s used by the RAF High Speed Flight to set a new world air-speed record (615.81 mph) on 7 September 1946, and Reaper was the alternative name for the Meteor GAF (Ground Attack Fighter) based on the F8; a private-venture demonstrator (G-AMCJ and later G-7-1) was flown in 1950, but the type failed to find a buyer.

A somewhat haphazard attempt to produce A VTOL Meteor conversion at RAF Kemble in the mid-fifties nearly had an unusual spin-off; although the severely denuded T7 (WF826) showed no inclination to get airborne in its new form, it displayed a remarkable talent for snow-clearing, and a plan was evolved to produce the conversion as the *Ground Mog*, until a lack of surplus airframes became apparent.

Meteor

The remarkably advanced **Meteor** four-seat, twin-engined tourer was conceived by sometime world land-speed record holder Sir Henry Segrave, and a prototype completed by Saunders Roe in 1930 was named the Saro-Segrave Meteor (G-AAXP). Responsibility was then passed to Blackburns, who developed a new metal fuselage in place of the former wooden structure and built two more

(G-ABFP and FR) which were renamed Segraves in recognition of Sir Henry, who had since been killed (13 June 1930) in an attempt on the world water-speed record.

Metropolitan

Although the **Metropolitan** name strictly belongs to only the Convair CV-440 twin-engined piston propliner, it was often used indiscriminately also for its predecessors the CV-240 and the CV-340, particularly in Europe. Continental Airlines had used the title Skystreamer for its CV-240s, but generally the civil Convair-Liners had been unnamed until the CV-440 **Metropolitan** was introduced in 1955. US military variants included the USAF C-131 and the Navy R3Y Samaritan transports, and the unnamed USAF T-29 navigation trainer.

The title of Cosmopolitan first appeared as a marketing name for the CV-540 conversion (CV-440s with Napier Eland turboprops) as operated by Allegheny Airlines, but became better known for the ten similarly powered CL-66B versions built by Canadair from 1960; these were known to the RCAF as the CC-109, and eight were re-engined with Allisons in 1966. America's first turboprop transport though was the prototype CV-240 converted in 1950 to take Allison 501-A4s as the Turbo-Liner, and although this proved impractical the more mature 501-D13 engine was later used in no less than 166 conversions (sub-contracted to Pacific Airmotive) of CV-340s and CV-440s to new CV-580 standard. The Super 580 is a further upgrade of this with 501-D22G engines (in a similar installation to that of the Lockheed P-3 Orion) and in 1985 Allison also

proposed a stretched Flagship or Turbo Flagship conversion, with a fuselage 17ft longer providing seating for up to 76 passengers. The **Flagship** never flew, but the same powerplant and stretch feature on the Model 5800 by Kelowna Flightcraft (KFC) of Canada first flew in February 1992.

Convair themselves devised the series of conversions powered by RR Dart turboprops, and supplied conversion kits to a number of operators in the sixties; converted CV-240s were re-designated CV-600s with the British engine, and 340s and 440s became CV-640s.

Mighty Mite

Old US Navy term for a small but powerful aircraft, such as the Grumman F8F Bearcat and the North American FJ-2/3/4 Fury. Less complimentary small-plane names were *Scooter* and *Tinker Toy*.

Migmaster

Noted for its variable-incidence wing, the US Navy's Chance Vought F-8 (F8U) Crusader was credited with nineteen MiG kills in Vietnam, and was also known as *The Last of the Gunfighters* by its crews, who boasted: 'when you're out of F-8s, you're out of fighters.' A single two-seat F8U-1T *Twosader* was built, partly in a vain bid to tempt the Royal Navy, and spent most of its life at the US Naval Test Pilots School, Pax River, where it was popularly known as the *Golden Football*. The last bastion of the *'Sader* is the French Aeronavale, where it is rightly cherished as *Le Crouze*.

Milan

Dassault themselves were the first to fly a canard version of the Mirage III/V with their 1970 **Milan (Kite)** which was essentially a Mirage V (later restyled Mirage 5) with controllable 'moustache' foreplanes on the nose, and a nav-attack system similar to the Jaguar. The Company tried again in 1982 with the Mirage IIING (Nouvelle Generation) incorporating fixed canards on the intake trunking, but it was not until upgrades became the vogue later in the decade that the world's Mirage fleets really began sprouting canards to improve manoeuvrability and take-off characteristics.

Israel Aircraft Industries gained considerable experience of the Mirage in the development of the **Kfir**, and also participated in South Africa's Cheetah programme, which as well as canards has a new dog-tooth to the wing leading edge, a small fence replacing the wing slot of the original Mirage IIIEZ/RZ, new avionics, and a lengthened Kfir-style nose; the ENAER Pantera (Panther) upgrade of Chile's Mirage 50s is basically similar, but lacks the wing modification. Other, unnamed upgrades including canards of various sizes have included the Dassault modification to Brazilian Mirage IIIs, IAI's work on Colombian Mirage 5s, Switzerland's independent programme, and Belgium's Mirage System Improvement Programme (MirSIP). Mara (a Patagonian rodent) is the name applied by Argentina to ten ex-Peruvian Mirage 5Ps fitted with an IAI Dagger-type nose to produce a ground attack aircraft similar to the Finger (upgraded Dagger; cf **Kfir**).

Mili-Trainer
Few lightplanes can have had such a cosmopolitan career as those which stemmed from Bjorn Andreasson's tiny BA-7 two-seater, distinguished by its slightly forward-swept wings. The

original BA-7 flew in America in October 1958 but was put into production in Sweden as the MFI-9 Junior by A.B. Malmo Flygindustri, who in 1962 also granted a licence for Bolkow to build it in West Germany as the Bo 208 Junior. The type was never one for the claustrophobic, so in November 1964 Bolkow introduced the Lange Junior (Long Junior) with a cranked spar to allow more leg room; a similar modification plus new elbow bulges also appeared on the Swedish-built MFI-9B and lightly armed **Mili-Trainer**. It was claimed that five of the latter, sometimes referred to locally as the Mini-COIN but better known internationally as the *Biafran Babies*, destroyed eleven Nigerian Ilyushin Il-28s and MiG-17s on the ground during the 1969 struggle for independence, flown by a group led by the 60-year-old Swedish mercenary Count Carl Gustav von Rosen.

, In 1969 Saab (who had taken over MFI the previous year) flew the enlarged MFI-15 Safari, with an optional third seat, T-tail, and double the engine power at 200hp; an armed version, sold to Pakistan and Denmark, was originally designated MFI-17 Supporter, but around 1979 this became the Safari as well. Pakistan received a total of 92 in kit form from Sweden between 1975 and 1982, known locally as the Mushshak (Proficient or Expert), and the Pakistan Aeronautical Complex (PAC) at Kamra also built over 100 from scratch. The same organisation has also developed the Shabaaz (King Eagle) which is better suited to local conditions with a more powerful 210hp Continental engine and improved air-conditioning; testing began in 1988, and the majority of Army and Air Force Mushshaks are expected to be brought up to the new standard.

Millipede

Also known as the *Tatzelwurm* (a
mythological Alpine dragon) the
German Arado Ar 232 transport had
a conventional-looking main
undercarriage, plus eleven pairs of
small idler wheels arranged down the
centreline of the fuselage like a
millipede's legs; on the ground the
main gear would 'kneel' to rest the
aircraft on the idler wheels, lowering
the floor of the freight hold to truck
bed height, and also allowing low-
speed taxiing over rough ground.

Miltrainer

Company name for the Valmet L-70
piston-engined trainer, better known
by its Finnish Air Force title of Vinka
(Blast or Arctic Wind). In its early days
the type was also known as the Leko-
70, but this was merely an abbreviation
of Lentokone, the Finnish word for
Aeroplane. The turboprop outgrowth
of the Miltrainer/Vinka is the L-90TP
Redigo.

Misteltoe

Well-known generic term for the
WW2 German weapon consisting of
(usually) a time-expired bomber
converted into a flying bomb, and
guided to its target by a fighter which
was carried piggyback-style until final
separation. So named after the
parasitic nature of the Misteltoe plant,
the term actually referred to the lower
aircraft only, but the combination (i.e.
both aircraft) was also known as
Father & Son after a contemporary
newspaper cartoon strip. The only
version to be used in anger was the
original Mistel 1 combination of Ju
88A-4 and Bf 109F, first used to attack
shipping in June 1944. Various Ju 88/
Fw 190 combinations were also tested
as the Mistel 2 and 3, and a Ju 88/Ta
152 scheme (apparently undesignated)

was flight-tested as late as April 1945.
Unbuilt projects included a Ju 287 jet
bomber with an Me 262 (Mistel 4) and
the purpose-designed Ju 268 jet-
propelled bomb coupled to a He 162
jet fighter (Mistel 5).

Mixmaster

Occasional USAF nickname, along
with *Push-Me Pull-You*, for the Cessna
O-2 Forward Air Control aircraft
(based on the civil model 337
Skymaster) with its unusual layout of
both tractor and pusher engines, the
latter made possible by a twin-boom
tail unit. An armed version was also
produced by French licensee Reims
Aviation under the original title of
Milirole, although the Zimbabwe Air
Force operates it as the Lynx. The O2-
337 Sentry by Summit Aviation of
Delaware is an uprated and armed
T337 Pressurised Skymaster
conversion which was introduced in
1980 and sold to a handful of Third
World countries.

   Like the much larger Fairchild
Dollar 19 the twin-boom layout has
allowed a number of interesting
conversions. In February 1983,
Spectrum Aviation of Van Nuys flew
their SA-550 Spectrum-One utility
aircraft, produced by removing the
forward engine, extending the nose by
4ft, and replacing the rear piston
engine with a PT6A-27 turboprop.
Broadly similar is Basler's Turbo 34
(PT6A-34) unveiled in the summer of
1993 with conversion slated to take
place in Thailand, but the earlier Brico
0-2 project featured an Allison 250
turboprop driving a ducted fan and a
nose kept free for sensors or miniguns.

   The most radical of all Cessna 337
conversions though was the 1973 AVE
Mizar, in which the wings, tail and
rear motor were mated with a Ford
Pinto car; after landing the aircraft

parts could be detached and left on a telescopic stand, allowing the car to drive away independently. Versions were also planned with Pontiac or Chevrolet cars, but the project came to a tragic end with the fatal crash of the prototype in 1973.

Mohawk
RAF-only name for the Curtiss Hawk 75A radial-engined fighters (the unnamed P-36 in the USAAF) diverted from French orders in 1940. Similarly, the Curtiss Hawk 81 development with an Allison inline motor was named Tomahawk (*Tommo* or the *Old War Hatchet*) in RAF service; British models equated to the USAAF's unnamed P-40 to P-40C. The further improved Hawk 87 series, recognisable by a deeper engine cowling and overall less 'hunched' appearance, first entered USAAF service as the P-40D, but as only 22 of that particular model were actually delivered, the new American service name of Warhawk was only introduced with the P-40E; RAF variants were Kittyhawks or *Kittys*, and equated to the P-40D to P-40N models. The lightweight P-40L was a stripped-down model, appropriately nicknamed *Gipsy Rose Lee*.

Mohawk 298
American conversion of the sixties-vintage Nord 262 (the Fregate [Frigate] in its N262C and D forms) French feederliner to comply with Federal Aviation Regulations Part 298. Nine Allegheny Airlines aircraft were converted by Frakes Aviation of Texas in the mid-seventies, and improvements included P&WC PT6A-45 turboprops in place of the earlier Turbomeca Bastans, new wingtips, new avionics, and improved cabin furnishings.

Monkey Killer
Withering term within the US fighter-bomber community, along with *Toothpick Maker*, for the Boeing B-52s used to wreak havoc on vast swathes of jungle (and hopefully something hiding inside it) in Vietnam. Otherwise, the mighty Stratofortress is surprisingly short of original nicknames; *Beast* and *Green Lizard* were rarely used, and even the more popular *Buff* is shared with a number of other types. Colossal Guppy was the apt title used by Aero Spacelines for a proposed freighter conversion with twelve engines, a 40ft diameter fuselage and a 200,000lb payload; a typical projected load was no less than eighteen Bell UH-1 helicopters.

Moonbat
Moonbat, *Bat* and *Bomber Destroyer* were all unofficial titles for the one-off XP-67 twin-engined fighter of 1944, the first indigenous design built by the fledgling McDonnell Aircraft Corporation.

Morko
Also known as the Morko-Morane, the Finnish Air Force's **Morko** (Ghost) was an adaptation of the French-built Morane-Saulnier MS 406 and MS 410 single-seat fighters with Soviet Klimov M-105P engines, made available from stocks captured by Germany. The first of roughly 40 conversions flew in February 1943.

Moskva
Moskva (Moscow) was the former type name of the Ilyushin Il-18 (NATO – Coot) four-turboprop airliner. A meteorological research variant is known as the Tsiklon (Cyclone).

Moskvitch

Named after the Soviet car, the **Moskvitch** was an improved version of the Mil Mi-1 (NATO – Hare) light helicopter with all-metal rotor blades, hydraulic controls and improved instrumentation, developed for Aeroflot in 1961. Although the name appears to have been dropped at an early stage, it is sometimes used erroneously in the West as a type name for all Mi-1s.

Mossie

A famous nickname for a famous aeroplane, *Mossie* (usually pronounced 'Mozzie') was of course shorthand for the de Havilland DH 98 Mosquito. As a nickname, *Wooden Wonder* (there was also *Balsa Bomber*) runs it a close second, and whilst it was not the only type so described, it was perhaps the most sincere application. Part of the Mosquito legend is the early disbelief shown in official circles to the concept of an unarmed wooden bomber, and as such the early title of *Freeman's Folly* can now be seen as a tribute to the tenacity of the Air Council's Sir Wilfred Freeman in championing the project. The anti-shipping Mk XVIII of 1943 (27 built) was also known as the Tsetse, possibly because of the bigger 'bite' from its nose-mounted 57mm Molins gun.

Mother

Slightly obscure and wholly unofficial name sometimes applied to the Martinsyde F3 experimental British fighter of 1917. Possible reasons are its large 'motherly' fuselage, a reference to the WW1 Mother tanks, the fact that it gave birth to the outstanding F4 Buzzard, or because shortly after it flew Martinsyde was allocated the initial letters 'Ma' for the names of

future products (see Appendix 3 for other companies' letters). Despite a marked superiority over the rival Sopwith Snipe, the Buzzard never really became operational with the RFC or RAF, but a number were sold after the Armistice to foreign air arms or converted for civil use; among the latter was the one-off Semiquaver racer (G-EAPX) fitted at one stage with short-span biplane wings, and later becoming a monoplane with the suspect 'Alula' wooden wing. Surplus RAF Buzzards were also sold by the Handley Page-controlled Aircraft Disposal Company, hence the generic title of *Disposalsydes*. Sadly, the innovative Martinsyde company itself, formed in 1908 as Martin & Handasyde, went into liquidation in 1921.

Motor Glider

Colloquial term for the Polish PZL-Mielec TS-11 Iskra (Spark) jet trainer, referring to its unusually good engine-out handling characteristics.

Mountaineer

Successful and relatively straightforward seventies-vintage rebuild of the Cessna L-19/O-1 Bird Dog, primarily for the civil market, by Ector Aircraft of Odessa, Texas. The standard machine retained the Bird Dog's 213hp Continental, but a Super Mountaineer was also available with a 240hp Lycoming 0-540.

Mouse

Though not formally a member of the Sopwith 'menagerie', the one-off *Mouse* biplane devised by F/Lt John Alcock and used by two wing RNAS at Mudros, was essentially a hybrid of other Sopwith types, taking its lower wing and forward fuselage from the Triplane, its tailplane from the *Camel*, and its upper wing (with modified ailerons) from the *Pup*. Also referred to as the Alcock Scout, the little biplane handled well, but was never actually flown by its creator, Alcock himself (later of Alcock & Brown fame) being taken prisoner before it was completed in 1917.

Muchacho
Spanish Air Force nickname (translated variously as 'servant', 'maid', or 'houseboy') for the long-serving de Havilland Canada DHC-4 Caribou STOL transport. In the RAAF the DHC-4 was also referred to as the *Wallaby*.

Mud Mover
Particularly popular with fighter pilots, *Mud Mover* is the standard put-down for ground-attack aircraft, particularly close-support types like the BAe Harrier and Fairchild A-10. In the eighties the term transmuted into the more respectable *Mud Fighter*, associated with a proposed new generation of low-cost tactical aircraft, i.e. the Rutan 151 ARES (Agile Response Effective Support) prototype of 1990, and BAe's unbuilt SABA (Small Agile Battlefield Aircraft).

Musketeer
Beech's Model 23 **Musketeer** lightplane was sometimes dubbed the 'Beechcraft **Cherokee**', and like its Piper rival Beech subsequently dropped the generic or family name in favour of individual titles; thus in 1972 the Musketeer Super R was renamed Sierra, the Musketeer Custom became the Sundowner, and the Musketeer Sport became simply the Sport. The Musketeer name lived on in the Canadian Armed Forces however, where in 1981 the CT-134A

Musketeer II (Sundowner) trainer replaced the earlier CT-134 Musketeer (Sport).

Nachtigall
Nachtigall (Nightingale) was the name allotted to the projected Arado Ar 234P night-fighter variant of the Luftwaffe's Ar 234B **Blitz** (Lightning), the world's first purpose-designed jet bomber.

Nancy
Play on the designation of the Curtiss NC flying boats, developed jointly with the Naval Aircraft Factory (hence NC for Navy-Curtiss) and famous for making the first (albeit staged) aerial crossing of the Atlantic in May 1919.

Nanok
A three-engined derivative of the RAF's Supermarine Southampton flying boat, the **Nanok** (Ice Bear) was rejected by Denmark in 1928 when the sole aircraft failed to meet performance targets. Later it was converted into an 'air yacht' for the Hon A.E. Guiness as the Supermarine Solent (G-AAAB).

Nightingale
Appropriate name for the USAF's C-9A aeromedical evacuation version of the McDonnell Douglas DC-9-30 commercial airliner; the first of 21 entered service in the summer of 1968, and there were also three VC-9C VIP transports. The Navy and Marine Corps named their C-9B transport counterpart the Skytrain II after the wartime R4D, and as well as seventeen new-build aircraft a further twelve DC-9s were purchased from commercial operators.

Ninak
Once likened to a DH 4 that had been 'interfered with' the Airco DH 9

93

bomber addressed the crew-position drawback of the so-called *Flaming Coffin* but was plagued by problems with its Siddeley Puma engine, installed German-style with exposed cylinder heads. Nevertheless, as a result of the 1917 RFC expansion plan something like 4,000 were ultimately built, and among numerous conversions were the long-serving South African Air Force DH 9J M'Pala with a Bristol Jupiter radial engine, and the one-off Wolseley Viper-powered Mantis. Much more successful than the original DH 9 was the Westland-developed DH 9A, with an American 400hp Liberty engine restoring the blunt-nosed look of the DH 4; the new mark soon became known as the *Ninak* (from the phonetic Nine-Ack) by which name it was fondly known throughout its long career with the inter-war RAF. The Armstrong Whitworth Tadpole of 1920 was a prototype three-seat naval spotter derivative with a deepened rear fuselage to accommodate a third crew member, which was put into production by Westland as the grotesque Walrus, festooned with a hydrovane, arrester gear, flotation bags, and all manner of external plumbing and bracing. More recognisable as *Ninak* variants were the pair of DH 15 Gazelles built in 1919 as testbeds for the 500hp Galloway Atlantic engine, and the one-off DH 9AJ Stag of 1926 with Bristol Jupiter engine and improved undercarriage.

Nomad
RAF name for the Northrop A-17A single-engined light bomber; 60 were received in 1940 and passed on to the South African Air Force.

North Star
Originally intended as just an individual name for the prototype, North Star came to be used as a type name for Canadair's Merlin-engined Douglas DC-4/C-54 Skymaster derivative (with some DC-6 features), although in service with Trans-Canada, Canadian Pacific and the RCAF the excessive cabin noise (only partly cured by an elaborate crossover exhaust modification) meant that it was frequently referred to as the *Noisy Star* instead. Twenty-two DC-4Ms were also delivered to BOAC, serving the British carrier faithfully throughout the fifties as their well-known Argonaut.

Rescuemaster is the all-but-forgotten name of the USAF's SC-54D air-sea rescue version of the standard P&W Twin Wasp-powered Skymaster (actually converted by Convair), but the most extensive of all DC-4 conversions must be the Aviation Traders ATL 98 Carvair (a name derived from 'Car-via-Air'). Allegedly conceived by managing director Freddie Laker in his bath, the Carvair had the flight-deck repositioned high above the hold, into which five cars could be loaded via a sideways-hinged nose door, the whole structure adding 8ft to the DC-4's length; a DC-7 type tail and DC-6 brakes were fitted, and despite the bulbous new front end, the cruising speed suffered by just four knots. A total of 21 conversions were carried out at Stansted up to 1968, serving with such operators as British United, Aer Lingus and BAF, before the hovercraft began to take over car ferry operations.

Nymph
The Norman Aircraft Company NAC 1 Freelance four-seat high-wing lightplane of 1988 first appeared back in 1969 as the one-off Britten

Norman BN-3 Nymph (G-AXFB).

Omega
Turboprop derivative of the Armée de
l'Air's Aerospatiale (Socata) TB-30
Epsilon trainer, with a Turbomeca
TP319-1A2 Arrius motor in place of
the usual Lycoming piston unit. A
private venture prototype (F-WOMG)
was flown in April 1989, but at the
time of writing none have been sold.

One-O-Wonder
Like many another Cold War warrior,
the McDonnell F-101 Voodoo was
blooded in the tactical heat of south-
east Asia, where the tremendous
power and range of the RF-101C
proved invaluable in the tactical recce
role. With the proud motto 'Alone,
Unarmed and Unafraid', the *Long
Birds* and their pilots who took their
turn 'in the barrel' (like the proverbial
fish) deep into North Vietnam were
both regarded as a breed apart.

One Shot Lighter
American nickname for early models
of the Mitsubishi G4M (Betty) twin-
engined bomber, which with virtually
no armour protection for either the
crew or fuel tanks were notoriously
vulnerable to gunfire. The Japanese
themselves apparently dubbed it the
Flying Lighter for the same reason,
although earlier in its career it had
been known as the *Flying Cigar*
because of its fat, rounded fuselage.

Flying Orange Box
Nickname derived from the boxy
biplane tail unit of the Handley Page
HP 42, the majestic four-engined
biplane airliner that so epitomised
Imperial Airways in the thirties.
Flying Banana came from the slightly
upswept rear fuselage. Although the
HP 42, as a type, was often referred

to as the Heracles, that was actually
the name of G-AAXC, the flagship of
the HP 42W Western Class fleet (G-
AAGX Hannibal was flagship of the
HP 42E Eastern Class).

Orel
Proposed 48-seat civil airliner
derivative of the Junkers Ju 290 four-
engined bomber/maritime recce type,
to have been built by Letov of
Czechoslovakia. A single L290 **Orel**
(Eagle) was assembled in 1946 from
German-manufactured components,
but no buyers were found and it was
eventually scrapped in 1956.

Orione
The MC205N **Orione** (Orion) was a
somewhat refined version of the
standard Macchi MC205 Veltro
(Greyhound) Italian fighter, differing
in its extended wings, aerodynamically
refined fuselage, and revised
armament; two prototypes were built
in 1942/43, but no production was
undertaken.

Osage
US Army name for its TH-55 primary
training helicopter, a variant of the
civil Hughes Model 269C; a total of
792 TH-55s were procured between
1964 and 1969, and the type was
finally retired from Army service in
1988. The improved Model 300 was
also made available from 1971 in Sky
Knight police patrol form with a
public address system, searchlight,
and underseat fibreglass armour
protection. Hughes (now McDonnell
Douglas Helicopters) switched
production of the 269/300 line to
Schweizer in 1983, and three years
later the latter company bought the
programme outright.

Osprey
CP-3C **Osprey** was the intended title

of the Canadian Armed Forces'
Lockheed P-3 Orion maritime patrol
aircraft, but both the name and the
designation were dropped following
the cancellation of the original order
in May 1976; barely two months later
however the order was reinstated, and
deliveries of eighteen began in 1980
with the title of CP-140 Aurora. The
basic Canadian aircraft have an
avionics suite similar to that of the
Lockheed S-3 Viking, but there are
also a trio of simpler CP-140A
Arcturus models in service for fishery
protection and environmental patrol.

Aries is the title of the US Navy's
dozen EP-3E electronic surveillance/
countermeasures conversions of P-
3As and EP-3Bs (Aries I) and P-3Cs
(Aries II), which are identifiable by a
belly radome and dorsal and ventral
'canoe' fairings. So far there have
been no military takers for the P-3
AEW&C (Airborne Early Warning
& Control) variant with a large dorsal-
mounted rotodome (E-3 Sentry-style),
but the US Customs Service has four
essentially similar aircraft for
detecting drug-running aircraft in the
Caribbean and Latin America; these
were given the formal title of *Blue
Sentinel* and the nickname *Blue
Canoe* when the first (a converted ex-
RAAF P-3B) entered service in June
1988, but neither title appears to have
survived a new low-viz paint job. For
the interception and close-in tracking
of suspect aircraft the Customs Service
also has a similar number of UP-3As
with nose-mounted Hughes AN/APG-
63 radar (similar to that in the F-15
Eagle fighter) which, lacking the
rotodome of the P-3 AEW&C, are
known as *Slicks*.

The basic P-3 Orion was of course a
derivative of the L-188 Electra
turboprop airliner, and Argentina has
modified at least three of these with

belly-mounted APS-705 search radar
as the L-188E Electron; the name may
appear to be an amalgam of Electra
and Orion, but actually comes from
Electra de Exploracion, after the
Escuadrilla Aeronaval de Exploracion
which operates them.

Owashi
Japanese Maritime Self-Defence
Force title (translating as 'Giant
Eagle') for the Lockheed P2V-7 (P-
2H) Neptune maritime patrol aircraft,
sixteen of which were supplied from
the US and a further 48 assembled
locally by Kawasaki. In 1988,
firefighting specialists Aero Union of
Chico, California put into service their
first SP-2H Firestar, a P2V-7
conversion with the usual
Westinghouse J34 auxiliary jets
deleted, a 7,570 litre retardant tank,
and a computer-controlled dispersal
system.

Ox-Box
Standard RAF-speak for the wartime
Airspeed Oxford twin-engined
trainer, a derivative of the civilian
AS6 Envoy light transport. The Envoy
was designed by N.S. Norway
(perhaps better known as the novelist
Nevil Shute) and A. Hessell Tiltman,
and in addition to 50 standard
machines there was also a single AS8
Viceroy (G-ACMU) with a cabin fuel
tank and deleted passenger windows,
built for the 1934 MacRobertson
England-Australia race; Mitsubishi
also built either ten or eleven Envoys
under licence as the Hina-Zuru
(Young Crane). Post-war, Airspeed
had considerable success with the
AS65 Consul, around 150 of which
were converted from surplus Oxford
airframes for charter and air taxi
work.

Packing Case Bomber
Rather like one of those early plastic model kits scaled to fit the box, the Short Stirling bomber was irredeemably compromised (though not to the extent sometimes claimed) by the requirement that it should fit the existing RAF hangars of the day, and even (in knocked-down form) standard Air Ministry packing cases. Also known as the *Flying Solenoid* because of its novel electrically-actuated undercarriage, bomb doors and flaps, the Stirling was not lacking in character however, and as the RAF's first four-engined heavy of the war was known to many of its devotees as the *Pulveriser*. The S37 *Silver Stirling* (PJ958) of 1945 was an abortive attempt to develop a 30-seat airliner conversion, after the fashion of the Lancastrian and Halton.

Flying Pancake (i)
One of the more intriguing American fighter projects of WW2 was the almost circular Vought V-173, flown for the first time in November 1942, and dubbed *Flying Pancake*, *Flapjack*, and *Zimmer Skimmer* after designer Charles H. Zimmerman. The V-173 itself was purely experimental, but was to have been followed by the XF5U-1 naval fighter (given the same nicknames) which was estimated to have a top speed of 500 mph, a landing speed of just 40 mph, and with its two huge props projecting from the front was perhaps even capable of VTOL operation. Gearbox problems continually delayed the XF5U-1 and with jets arriving on the post-war scene the programme was cancelled; the complete but unflown prototype was broken up in 1947, but the original V-173 was saved and presented to the Smithsonian Institute.

Flying Pancake (ii)
Occasional term for the US Navy's Grumman E-2 Hawkeye AEW aircraft, after its large, dorsal-mounted rotodome. In the case of the E-2 the more general term *Hummer* has also been attributed to the sounds of its Allison T56 turboprops or the mass of electronics it carries.

Panda
Name used in the mid-eighties, and apparently discarded later, for Western promotion of China's Harbin Yun-12, the stretched, turboprop-powered development of the roughly Islander-class Yun-11.

Paragon
Among names considered for the Handley Page (Reading) HPR3 Herald airliner in 1954 were (in nothing more than alphabetical order) Bolivar, Consort, Continental, Crusader, Diplomat, Hamilton, Haulier, Paragon, Paramount, Pilgrim, Profit, and Sovereign. The two Herald prototypes (G-AODE and -AODF) were built with four Alvis Leonides piston engines, but the type went into production with twin Rolls-Royce Dart turboprops as the HPR7 Dart Herald.

Partizan
Yugoslavian Air Force name for the Aerospatiale SA 341H Gazelle helicopter, 21 of which were delivered from France with a further 132 built by Soko at Mostar. Soko also built the SA 342L model, and from it they derived the anti-armour Gama (from Gazela Maljutka) with four AT-3 Maljutka (NATO – Sagger) wire-guided missiles, and the Hera (from Helicopter Radio) recce model.
    A pair of SA 341G Gazelles (N51BT and N52BT) were converted

by Wright Airlift International into fictional Blue Thunder helicopter gunships for the Columbia film and TV series of the same name.

Peacemaker
The USAF never formally recognised the Convair title (selected in a 1949 competition) of **Peacemaker** for the B-36, but as the world's biggest-ever bomber it would never be short of alternative names; nor could the sight and sound of a B-36 ever be forgotten, particularly the B-36D and subsequent 'six turning, four burning' models, whose unique combination of four jets (podded in pairs) and six pusher props produced such a distinctive 'beat'. A wing area of 4,772 sq ft (span was 230ft) earned such titles as *Magnesium Monster* (or-*Cloud*) and *Aluminium Overcast*; *Flying Cigar* referred of course to the 163ft-long cylindrical fuselage, and with an early reputation for last-minute malfunctions the B-36 almost made the *Ramp Rooster* title its own. As one of the mainstays of America's nuclear deterrent in the fifties, the B-36 was sometimes referred to as the *Big Stick*, after Theodore Roosevelt's maxim, 'Speak softly and carry a big stick'. Crusader was the title given to the single NB-36H (51-5712) with a nuclear reactor in the tail, not for propulsion, but to test the effects of radioactivity on aircraft systems.

Peanut Special
Contemporary British nickname for the tubby Grumman F4F Wildcat naval fighter, known more formally to the Royal Navy until early 1944 as the **Martlet**. The vast majority of Wildcats (well over 5,000) were actually built by Eastern Aircraft (General Motors) as FM-1s and FM-2s, and at one stage they hoped to rename their version

the Mongoose (cf *Chuff*). The F4F-3S *Wildcatfish* was a single aircraft fitted experimentally in February 1943 with a pair of Edo floats.

Peashooter
Famous nickname for the little Boeing P-26 fighter monoplane of the early thirties, supposedly because of the long gun extension-tubes projecting between the cylinders of the radial engine.

Pedro (i)
The first German bomber unit in the Spanish Civil War (flying Ju 52/3ms) was the Kampfstaffel Moreau, who jocularly dubbed themselves the 'Pedros & Pablos', names which were later taken up unofficially for the Condor Legion's Heinkel He 111 and Dornier Do 17 *Flying Pencil* bombers respectively.

Pedro (ii)
Affectionate name (originating as a callsign) during the Vietnam war for the Kaman H-43 Huskie rescue helicopter, also known as the *Whistling Outhouse* after its square, solid appearance. Overshadowed in later years by the *Jolly Green Giant*, the Huskie was one of the unsung heroes of the south-east Asia conflict, and was credited with the quickest 'save' of all, rescuing an OV-10 Bronco pilot just 92 seconds after ejecting on 23 November 1972.

Pembroke
Although the RAF's Pembroke C1 communications aircraft is perhaps the best remembered of the family descended from the civil Percival P50 Prince, the Royal Navy was both the first and biggest customer. Starting in 1950 the Senior Service took delivery of four short-nosed Sea Prince C1

comms aircraft, 42 radar-equipped T1 navigation trainers, and three C2 comms aircraft externally similar to the trainers with their extended noses. Pembroke production for the RAF amounted to 44, and others were supplied to the air forces of Belgium, Denmark, Finland, West Germany, Rhodesia and Sweden. Perhaps the least known variant of the Prince was the President (originally Prince V) executive/VIP model, six of which were completed including one converted Pembroke. Percival's first high-wing twin was the one-off Gipsy-engined P48 Merganser of 1947, but despite a strong similarity to the Prince the latter was in fact a completely new design.

Flying Pencil
Having caused such a stir at the Zurich meeting of July 1937, the sleek Dornier Do 17 soon became famous in aviation circles as the *Flying Pencil* and in Britain also by the brand name of *Eversharp*; *Pregnant Pencils* were the variants with enlarged crew compartments, notably the Do 17Z and Do 215, and to a lesser degree the larger Do 217 series. The Condor Legion christened the Do 17 *Pablo* (cf *Pedro* for origins) and the more common civil war nicknames of *Haddock* and *Bacalao* (*Codfish*) are thought to allude to the thin strips into which the Spanish sometimes cut these fish. Kauz (Screech Owl) was the formal name of ten Do 17Z-3s converted into interim night-fighters in 1942, comprising a single Do 17Z-6 with a nose-cone similar to that of the Ju 88C-2, and nine definitive Do 17Z-10 Kauz IIs. In service with the Royal Bulgarian Air Force the Do 17M was known as the Uragan (Hurricane), and after the Soviet occupation actually flew against the Wehrmacht.

Penetrator
Had it flown and entered service with the US Navy as planned, the General Dynamics/McDonnell Douglas A-12 Avenger II would undoubtedly have remained one of the world's major warplanes well into the 21st century. From a nomenclature viewpoint it was unusual in being named after another company's product, and it was perhaps to forestall objections from Grumman (who had badly needed to be on board the project) that the 'II' was later dropped. Among other formal names considered were Penetrator, Enforcer, Ghost, Shadow, Seabat, and Stingray. The A-12 was cancelled suddenly in January 1991, and was later revealed as having been a pure triangle in plan-form, with a span of around 69ft and a length of approximately 36ft.

Penguin
In WW1, French pilots were often given their initial training on *Roleurs* or *Penguins*, 'aircraft' that could taxi and simulate take-off runs to give the pupil the feel of the real thing, but which because of their clipped wings and low power could not actually fly. Among the companies producing machines for this 'grass-cutting' technique were Blériot and Morane Saulnier.

Peshka
Common nickname for the Russian Pe-2 (NATO – Buck) high-speed light bomber of WW2, and essentially a diminutive form of Petlyakov, the head of the design bureau responsible.

Petrel
Although Percival did not formally recognise the name of Petrel suggested by *Aeroplane* editor C.G. Grey for their handsome Q6 twin, it

was later adopted by the RAF for its eight WW2 aircraft.

Pfiel
Willy Messerschmitt's career as an aircraft designer did not end in 1945, and in the early fifties he designed for Hispano the Me 200 Pfiel (Arrow) tandem-seat jet trainer, which went into production for the Spanish Air Force as the HA-200 Saeta (again translating as 'Arrow'). The HA-231 Guion development, proposed in side-by-side trainer and six-seat liaison forms, failed to materialise, but Egypt built the armed HA-200B Saeta under licence at the Helwan Air Works as the Al Kahira (The Cairo).

Picio
Spanish Air Force nickname for the indigenous CASA C-212 Aviocar STOL utility transport (military designation T12), after a particularly unattractive character in a Spanish fairytale.

Flying Pig (i)
Nickname originating within Instone Air Line for the singularly unattractive Vickers Vulcan airliner of the early twenties.

Flying Pig (ii)
Contrary to some accounts, it was not simply the plump outline of the RAF's Saro Lerwick (properly pronouced 'Lerrick') that earned the twin-engined flying boat this title, but also an unpleasant tendency to roll and yaw (i.e. to wallow) when flown hands-off.

Pika
The Australian-developed GAF Jindivik (Aborigine for 'The Hunted One') target drone has served the RAF and other air forces for many years, but perhaps less well-known

were the pair of manned prototypes built under the name **Pika** and first flown at Woomera in October 1957.

Pilot Maker
A classic term for a classic aircraft, *Pilot Maker* reflected not only the vast numbers of WW2 aircrew trained on the ubiquitous North American T-6 Texan/SNJ (*Jaybird*) family, but also the fact that you were only considered a 'real' pilot once you had mastered it. The type is also well-known for the noise from its high-speed ungeared propeller, which earned the nickname of *Old Growler* in America, whilst in Britain the *Pilot Maker* title was quickly punned as *Window Breaker*. Harvard was of course the name used by the RAF and most Commonwealth air forces, although the RCAF called its fixed-gear NA-64s (diverted from French contracts) Yales, and the variant built by Commonwealth Aircraft in Australia was known as the Wirraway (an Aboriginal word meaning 'Challenger'). Post-war, Commonwealth also produced a single-seat agricultural conversion of the Wirraway, named Ceres after the Roman goddess of agriculture. The South African Air Force's long-serving Harvards were sometimes referred to as *Cannibal Harvards* (spare parts being a problem for such an old type) and Mosquito was the name used for 98 armed LT-6G conversions used as USAF Forward Air Control (FAC) aircraft in Korea.

Pin
Popular RAF term for the useful Scottish Aviation Pioneer STOL lightplane, also known originally as the Prestwick Pioneer after its birthplace; the *Twin Pin* was of course the SAL Twin Pioneer, which despite the name had relatively little in

common with its forebear. Pegasus
and Clydesdale were titles associated
with the projected single-engined
Pioneer Mk 4 in the late fifties with a
boxy new fuselage seating up to nine.

Pinball
There was nothing wrong with being
the pilot of a Bell RP-63 Kingcobra –
as long as you didn't object to being
deliberately shot at by your own side.
No less than 332 of these target
aircraft were either built as new or
converted from standard P-63 fighters,
featuring over half a ton of toughened
skin (attacking aircraft were given
special frangible ammunition), bullet-
resistant canopies, and even a red
wingtip light which lit up like a pinball
machine when the RP-63 was hit.
Whilst clearly resembling its earlier
stablemate, the P-63 Kingcobra
should not be regarded as a variant of
the P-39 Airacobra (cf *Little Shaver*).

Pinocchio
Coming from Switzerland (the
assumed locale of the Disney story)
and with a long turboprop nose that
has to be seen to be believed, there
could only ever be one nickname for
the F&W C-3605 Schlepp target tug.

Pioneer (i)
Unofficial but appropriate name for
the little Gloster E28/39 which,
powered by Sir Frank Whittle's 860lb
thrust W1 engine, made Britain's and
the Allies' first jet flight at RAF
Cranwell on 15 May 1941. The
mysterious new aircraft was also
referred to as the *Squirt* or sometimes
the *Gyrone* and the name Tourist
appeared in contemporary
documents, but the precise status of
the latter is unclear. Two aircraft were
built, purely for research; W4046
crashed on 30 July 1943, but the

historic first aircraft, W4041, survived
and was put on permanent display in
the Science Museum in 1946.

Pioneer (ii)
N23 Pioneer was Northrop's
commercial title for what must be
their least remembered production
type, the USAF's YC-125 Raider
STOL transport; 23 of the high-
winged tri-motors were delivered from
1950, but used mainly as instructional
airframes.

Pippi
Affectionate West German Air Force
term for the Piaggio P149D trainer/
liaison aircraft, one of the earliest
types to be used by the reconstituted
Luftwaffe, with the first Italian-built
machine (others were built by Focke-
Wulf) being delivered in May 1957.

Pirate
Despite the fact that both names had
already been used by Chance Vought
for US Navy jets, **Pirate** and Cutlass
were both reportedly considered for
the navalised de Havilland DH 110
before settling on Sea Vixen. In plain
Vixen form the name had already
been earmarked for an RAF version
of the twin-boom, all-weather fighter
(passed over in favour of the Gloster
Javelin), and in service with the Royal
Navy from 1959 to 1972 the 'Sea'
prefix was usually dropped anyway in
everyday usage.

Plastic Chinook
Inevitable nickname for the 1987
Boeing Model 360 all-composite
research helicopter (N360BV) despite
having little in common with the larger
CH-47, except the tandem-rotor
configuration.

Plastic Parrot
Royal Australian Air Force term for
the pretty but unloved New Zealand
Aerospace CT-4 Airtrainer, infamous
for its backache-inducing cockpit.

Plastic V-Force
Not widely known are the Valiants,
Vikings and Vanguards operated by
the RAF in the eighties and nineties.
These are among the fleet of largely
glassfibre sailplanes operated with
special service names by the
Volunteer Gliding Schools, who
provide air experience for Air
Training Corps cadets. Types have
included the winch-launched Valiant
(Schleicher ASW-19), Viking (Grob
G103 Twin Acro) and Vanguard
(Schleicher ASK-21), and the self-
launching (i.e. powered) Venture
(Scheibe SF250B Motor-Falke, built
under licence by Slingsby as the T61E)
and Vigilant (Grob G109B).

Plywood Bullet
Occasional and highly appropriate
nickname for the Lockheed Vega, the
high-speed monoplane that did so
much to establish the Lockheed name
in the late twenties/early thirties; the
most famous of all the Vegas was of
course Wiley Post's 'Winnie Mae'.

Pocket Herk
Slightly unfair description of the
Franco-German Transall C160 tactical
transport; in some respects the
European type is more capable than
the Lockheed C-130 and actual
dimensions are remarkably similar,
although perhaps having only two
engines makes it appear otherwise.
First flown in 1963, the Transall
(Transporter-Allianz) C160 was
known in South Africa (at least) as the
Trannie, and Gabriel is the formal
name of a pair of Armée de l'Air

aircraft modified for the Sigint
(Signals Intelligence) role.

Pogo
Well-known nickname for the
extraordinary Convair XFY-1
experimental tail-sitting VTOL
fighter, designed to take off and land
by hanging on its nose-mounted, 16ft
contra-rotating props. Despite
winning acclaim (and the Harmon
Trophy) for the world's first non-
helicopter/autogyro vertical take-off
in 1954, flight tests revealed severe
handling difficulties (particularly in
view of its intended role as a
shipborne interceptor) and
development was abandoned after
about 40 hours of flying.

Polecat
In keeping with Grumman's fighting
feline nomenclature, this was the
nickname of the pair of X-29 research
aircraft built in the early eighties to
evaluate the Forward Swept Wing
(FSW) concept. To aid development,
the X-29 used the forward fuselage of
the Northrop F-5 and the main
undercarriage of the Lockheed F-16;
power came from a single General
Electric F404-GE-400 of 16,000lb
thrust.

Flying Porcupine
Big and beautiful it may have been,
but to hostile aircraft the RAF's Short
S25 Sunderland flying boat could also
be a surprisingly prickly customer,
eliciting from Luftwaffe pilots the
respectful title of *Fliegende
Stachelswein* – the *Flying Porcupine*.
Never was that prickliness better
illustrated than by the *Sun* which was
set upon over the Bay of Biscay by no
less than eight Ju 88s, and instead shot
down three of its tormentors.
Stickleback was the British nickname

for those aircraft (principally Mk IIs and Mk IIIs) with four ASV Mk II radar aerials along their spines.

The Sunderland had its origins in the pre-war **Empire Boats** and from December 1942, 24 Sunderland Mk IIIs with faired-over turrets and typically utilitarian wartime seating were also supplied to BOAC; after VE Day 22 of these were fitted out to peacetime standards as the Hythe class with accommodation for 24 passengers in daytime configuration and sixteen in night-time *Slumberland* form. A more extensive civil conversion, with streamlined nose and tail in place of the Hythe's blanked-off turrets, also appeared after the war as the S25 Sandringham; versions produced were the Mk 1 Himalaya class (one only, for BOAC), Mk 2 and Mk 3 Argentina class (three and two, Dodero Aero-Navigacion), Mk 4 Dominion, later Tasman class (four, Tasman Empire Airways Ltd), Mk 5 Plymouth class (nine, BOAC), Mk 6 Norwegian class (five, DNL) and Mk 7 Bermuda class (three, BOAC).

A separate line of development began in August 1944 with the appearance of the RAF's S45 Seaford, originally designated Sunderland Mk IV, and differing by having 1,800hp Bristol Hercules engines in place of the usual 'Peggys' (Pegasus) or P&W Twin Wasps, four-blade props, a taller curved fin, and revised hydrodynamics. Just two prototypes and eight production Seafords were completed as such, but post-war twelve new Solent 2 commercial derivatives were ordered on behalf of BOAC (Solent 1 was an unbuilt cabin version); six of the ex-RAF Seafords were later converted as Solent 3s, and four Solent 4s were also built for Tasman Empire Airways Ltd. BOAC ended its flying boat services in

November 1950, and the last-ever British flying boat operation ceased in September 1958 when Aquila Airways withdrew its Solent service to Funchal (Madeira).

Postal
Just as various American aircraft were given misleading names or designations for 'funding purposes' (e.g. the B-29D Superfortress becoming the 'new' B-50 in December 1945 to save it from post-VJ Day cancellations) so there was a tendency in Britain between the wars to justify new types by calling them **Postal** aircraft, and hinting at a role in the upkeep of the Empire. The best known example was the Fairey Long-Range Monoplane, with which it came to be almost an alternative type name, although the subterfuge was also partly to mask aspirations towards breaking the world long-distance record (ultimately achieved in February 1933).

Flying Postbag
Predictable nickname for the compact Boulton and Paul P64 Mailplane of 1933; the sole prototype (G-ABYK) crashed on its third flight, but a pair of refined P71As (G-ACOX and G-ACOY) were delivered to Imperial Airways in 1935 and operated as the Boadicea class.

Flying Potato
Unflattering nickname of the original Martin Marietta X-24A (alias SV-5P) lifting-body research aircraft (cf *Flying Stone*), which made its first powered flight in March 1970. Two years later the X-24A was rebuilt as the X-24B with a more triangular cross-section and long slender nose, in which form it became known a little more kindly as the *Flying Flatiron*.

Pot Belly
Title used by Vickers test pilot Harold
Barnwell for their one-off FB16D
scout of 1917. The name was recorded
in the classic *Flying Fury* by the great
British ace James McCudden, who
used the tubby little biplane as a
runabout whilst on leave from the
Western front.

Flying Powerstation
At the behest of the RLM, virtually
every system in the first two Focke-
Wulf Fw 191 prototypes was
electrically actuated, earning the
medium bomber the title of *Das
Fliegende Kraftwerk* or *Flying
Powerstation*.

Pregnant Guppy
Obvious nickname for the 1939
Consolidated Model 31 twin-engined
flying boat, with a length of 73ft and a
fuselage depth of no less than 22ft.
The P4Y Corregidor production
version was cancelled, but the
advanced high aspect-ratio, 110ft
Davis wing and the twin tail unit were
both adapted for the famous B-24
(Consolidated Model 32) Liberator
bomber.

Pregnant Moth
Nickname of the four-seat General
Aircraft Co. (Genairco) biplane,
derived from the famous de Havilland
DH 60 Moth and built at Sydney,
Australia from 1930.

Pregnant Spitfire
Also known as the *Sawn-off Spitfire*,
the RAF's feisty Boulton Paul Balliol
trainer of the fifties had a Rolls-Royce
Merlin 35 engine and a performance
not unlike that of the famous WW2
fighter.

Prog
Prog and *Perce* were both used for the
RAF's wartime Percival Proctor
trainer/communications aircraft, the
former borrowed from Oxford
University slang. Developed from the
pre-war Vega Gull tourer, the Proctor
I first flew in October 1939, and the
final RAF Mk IV version of 1943 was
initially to have been known as the
Preceptor. Peewit was the name
applied to a proposed post-war
development of the Mk V (civilianised
Mk IV) featuring metal construction
and new sliding doors, amongst other
improvements.

Prometheus
Although the Supermarine **Swift** was a
disappointment to the RAF, one of
the prototypes (VV119) had a brief
moment of fame when it appeared as
the new **Prometheus** fighter in the
1951 British film *The Sound Barrier*.
Directed by Alexander Korda and
scripted by Terence Rattigan, the film
provides an attractively atmospheric
image of British aviation in the
pioneering fifties, and as such perhaps
the notion of pushing through the
sound barrier by 'reversing the
controls' can be overlooked.

Provence
Air France class name for the
distinctly Gallic Breguet 763 Deux
Ponts (Double Deck) piston airliner,
seating 59 on the upper deck and 48
on the lower. Six of the Air France
fleet of twelve were converted into
Universel passenger-freighters in
1964/5 and used partly to support the
Concorde programme by carrying BS
Olympus engines to Toulouse; the
remainder were taken over by the
Armée de l'Air. Previously the air
force had taken delivery of four (out
of a planned fifteen) of its own

Breguet 765 Sahara version – easily distinguishable by its tip-tanks – and total production of the Deux Ponts series came to just twenty, including three development aircraft (also used by the air force) and the original Breguet 761 prototype of 1949.

Pulser
The latest known product of Lockheed's 'Skunk Works' is the mysterious Aurora strategic reconnaissance aircraft, allegedly capable of speeds in the order of 3,800 mph and built to succeed the SR-71 *Blackbird*. Power is thought to come from an advanced pulse-detonation wave engine, producing a distinctive hollow pulsating sound and a strange form of contrail that has been likened to 'doughnuts on a rope'. The Aurora title was first revealed in a supposed security slip-up in February 1985, but in time the name will probably be seen as authentic as 'F-19 Ghost' was in relation to the actual F-117 (cf *Wobblin' Goblin*).

Pumpkin
Nickname applied to the 1913 Vickers No. 26, an equal-span variant of the Henri Farman biplane, after its crude two-seat crew nacelle.

Pup
The much-loved Sopwith Pup fighter owes its name to Brig. Gen. (later AVM Sir) Sefton Brancker (1877–1930) who upon seeing it for the first time, parked alongside its larger stablemate at Brooklands in the spring of 1916, reputedly exclaimed 'My God, your 1½ Strutter has had a pup!' The name quickly caught on, but lesser men in higher places were aghast at such a frivolous title, and amid the horrors of the Great War made a minor career out of issuing

edicts that it was to be referred to as the Sopwith Type 9901 (RNAS), or simply Sopwith Scout (RFC). The only result of their efforts was that the name became so indissolubly linked with the little scout that any arguments over 'official' or 'unofficial' titles became somewhat irrelevant. Post-war, a two-seat sports version with slightly-swept wings was developed as the Dove, and the last of ten built (G-EBKY) was 'reverse-engineered' in 1937/38 to become the Shuttleworth Collection's Pup N5180.

Q-Bird
Archaic US Navy term (also *Queer Bird*) for electronic warfare aircraft, dating from the days of the old pre-1962 designation system which denoted such types with a 'Q' suffix, e.g. F3D-2Q Skyknight, AD-4Q Skyraider, etc.

Queen
British title associated with radio-controlled target drone aircraft in the thirties and forties. The original trio of Fairey Queens were derivatives of the well-known Fairey IIIF biplane, built in 1931 to assess the ability of Royal Navy ships to defend themselves against air attack, but the best known type must be the de Havilland DH 82B Queen Bee; despite the designation and a strong resemblance to the famous DH 82A Tiger Moth, the Queen Bee was closer structurally to the DH 60GIII Moth Major, and a total of 387 were built up to September 1944. Also purpose-built for the role was the Airspeed AS30 Queen Wasp (seven only; cf **Clay Pigeon**) but other types were adaptations, i.e. the Queen Seamew (30 conversions) variant of the Curtiss SO3C **Seamew**, and the Queen Martinet (66) version of the Miles

Martinet target tug (itself derived from the Master II trainer).

Queen Air
The popular Beech Model 65 **Queen Air** light piston transport originated in 1958 as a civil counterpart to the US Army's L-23F, which used the wings and tail of earlier L-23 models (cf *T-Bone*) but had a completely new, deeper fuselage. The Army name of Seminole carried over to the new L-23F (later U-8F) but **Queen Air**s in service as trainers and liaison aircraft with the Japanese Self-Defence Forces have the alternative name of Umibato (Sea Dove).

Queen Airliner was the name used for Models 79 and 89, which were optimised more for third-level operators than for corporate use. A stretch of the Model 65-80 in 1966 produced the Model 99 with seating for seventeen, a long baggage-carrying nose, and the P&WC PT6A turboprops of the King Air; the original Models 99 and 99A were unnamed, but the A99 and B99 variants were known as the Beech Airliner, and the heavier and more powerful C99 was the Commuter. At least six of the A99s supplied to the Chilean Air Force from 1982 were fitted locally by ENAER with ventral radomes and used as Petrel A maritime recce aircraft and Petrel B Elint platforms.

Queenaire
As well as the Excalibur conversion of the Twin Bonanza (cf *T-Bone*), Swearingen Aircraft also turned their attentions to its big sister, the Beech Model 65 **Queen Air**. Initially known as the Swearingen 800 or Queen Air 800, the conversion centred around new 400hp Lycoming piston engines, with new engine mounts and three-blade props, but also featured Excalibur modifications like the new undercarriage doors. Actual conversion work was sub-contracted to Excalibur Aviation, who in 1970 acquired the rights to the programme and and renamed it **Queenaire** 800 or 8800 (depending on the particular Beech model under conversion).

Swearingen, meanwhile, used the basic **Queen Air** wing and Twin Bonanza undercarriage in their new SA-26 light corporate aircraft, named Merlin to preserve the Arthurian theme established with the Excalibur; the original prototype had Lycoming piston engines, but production models, certificated in 1966, had P&WC PT6A (Merlin IIA) or Garrett TPE-331 (Merlin IIB) turboprops. The 1969 Merlin III broke the Queen Air link by introducing a completely new wing which, combined with a hefty 17ft fuselage stretch, subsequently produced the Merlin IV (corporate) and the Metro nineteen-seat airliner; the latter was developed in conjunction with Fairchild, who in 1971 gained control of Swearingen. The Merlin IV title continued to be used for corporate versions of subsequent Metro developments (i.e. the corporate version of the Metro III is the Merlin IVC) and the Metro is also available in cargo form as the Expediter. The USAF version of the Metro III is the C-26.

Queen of the Humber
Local name for the RAF's majestic Blackburn Iris biplane flying boat, assembled at Brough and launched there for the first time on 19 June 1926.

Quiet Giant
German nickname for the Airbus A300 twin-engined widebody airliner.

Eastern Air Lines, afraid of presenting a downmarket image of a flying bus service, assigned the name Whisperliner to its A300s. To replace the four ageing Guppy 201s (cf *Strat*) used to transport components between its partners, Airbus Industrie is to have a similar number of A300-600ST Super Transporters or *Super Flippers*, modified from new-build A300-600Rs by the associated Franco-German SATIC organisation. Apart from the new 53,000 cubic ft Guppy-style fuselage, changes include a lowered cockpit to allow front-loading, a new fin fillet, and new endplate fins.

Quirk

'They were so damn slow they wouldn't go, and they called them Raf 2cs' ran an old RFC song about the Royal Aircraft Factory BE2c, a major model of the observation type that was once lauded as *Stability Jane* and later universally condemned when that same docility rendered them so hopelessly vulnerable in action. More than any others, BE2 crews were regarded as 'Fokker Fodder' in 1915/16, and in 'Bloody April' 1917 the RFC lost at least 60 in action. Considering their reputation, *Quirk* seems surprisingly mild, but perhaps a clue may lie in its concurrent use also for novice or trainee pilots, thoroughly dangerous creatures also known as 'Huns'. A single BE2c was modified in 1915 with the engine moved aft slightly and a prominent new observer's nacelle mounted ahead of the propeller; this earned the resultant BE9 the nickname *Pulpit*, although the term was also shared with other similarly configured types, notably the French SPAD SA1 to 4, and single-seat SG1.

Rabbit

Nickname of the British late WW1-vintage Alliance P1 trainer, stemming from the fact that the design was begun by Ruffy, Arnell & Baumann as the RAB15; the type failed to live up to its prolific namesake, with only one (G-EAGK) produced.

Flying Ram

So strong and fast was the Northrop XP-79B flying wing (particularly the uncompleted rocket-powered XP-79) that the possibility of ramming enemy aircraft was allegedly considered, although Jack Northrop himself denied any such plans. Sadly, America's most exciting wartime fighter project came to a disastrous end when the sole jet-powered XP-79B crashed during its maiden flight on 12 September 1945.

Rana

Rana and Pratap are Indian Air Force names for, respectively, the Soviet Mil Mi-8 and improved Mi-17 (NATO – Hip) helicopters; Rana was the title of the dynasty founded in Nepal by Jung Bahadur, and Maharana Pratap was a famous warrior king of Rajputana.

Ranger

One of a spate of abortive maritime patrol projects in the seventies, the de Havilland Canada Dash 7R **Ranger** was based on the civil DHC-7 Dash 7 high-wing four-turboprop STOL airliner; a Litton LASR-2 radar was to have been mounted in a belly radome, but much of the internal equipment would have been palletised to allow the **Ranger** to be quickly converted for transport duties. None were sold as such, but in 1986 the Canadian Dept. of the Environment belatedly took delivery of a single Dash 7IR (Ice Reconnaissance) with a dorsal-mounted observation cabin, Canadian Astronautics SLAR on the port-side, and a Laser Profilometer for establishing the size and shape of icebergs. The Canadian Air Force operated a pair of Dash 7 transports as CC-132s, but in 1986/87 these were traded-in for CC-142s (Dash 8s).

Rat
Like many Soviet types of the day, the pugnacious little Polikarpov I-16 monoplane fighter flew with the Republican forces in the Spanish Civil War, becoming known to them as the *Mosca* or *Fly* (the I-16 Type 10 was the *Super Mosca*) and more famously by its opponents as the *Rata* or *Rat*; indeed, the latter was so well-known that it was later taken up by Luftwaffe pilots after the invasion of the Soviet Union in 1941. Also known in its homeland as the *Yastrebok* (*Little Hawk*) or *Ishak* (*Little Donkey*) the I-16 is historically important as the world's first in-service cantilever monoplane fighter with a retractable undercarriage, a fact which Western aviation circles were either ignorant of or reluctant to concede. During the Spanish war it was frequently labelled a 'Boeing' in the haphazard assumption that the Republicans were operating P-26s or a Soviet copy of it, and as late as mid-1941 the *Aeroplane Spotter* was still blithely dismissing it as a copy of either the P-26 or Gee Bee Sportster, no less than six full years after the I-16 had made its public debut at the 1935 May Day flypast.

Raven
Unusual among US Army nomenclature for not being an Indian name, **Raven** was the title given to the military H-23 version of the commercial Hiller 360 light helicopter, also known in civil guise as the UH-12

(from United Helicopters). Amongst military export customers were the Royal Navy (21 unnamed Hiller HT2 trainers) and Canada, which designated its 24 OH-23G variants as CH-113 **Nomads**. Phoenix was the title of a comparatively obscure mid-eighties conversion by Craig Helicopters Inc., replacing the Lycoming VO-540-C2A engine of the standard Hiller 12E with a turbo-super-charged TIVO-540-A2A. UH-12E production was restarted by Rogerson Hiller Corporation of Port Angeles, Washington in 1984 with the new name of Hauler.

Flying Razor
RAF term for the Germans' late-WW1 Fokker DVIII scout, after its unusually small parasol wing.

Razor
The Soviet Union's Polikarpov R-5 was not only one of the more effective of the general-purpose biplanes that appeared in profusion between the wars, but also one of the most prolific, with roughly 7,000 built following its first flight in 1928. Over 30 served with the Republicans in the Spanish Civil War, where the type was known as the *Rasante*; translated variously as *Razor* or *Scraper* the name was probably inherited from the Soviet Union, where ground-strafing was commonly known as 'shaving' (cf *Little Shaver*). The improved R-Z (also known as the S-Z) also served in Spain, and whilst it has been claimed that this alone was known there as the *Natacha* it seems unlikely that such a clear distinction would have been made between two such closely related types.

Red Dog
Popular nickname for a pair of Lockheed Orion 9B six-seat low-wing

monoplanes which, with a vivid red
colour scheme, served with Swissair
from 1932 to 1935.

Reichenberg
The Nazis called it the
Vergultungswaffe Ein (V-1) or
Reprisal Weapon One, and indeed
there can have been few more chilling
sounds in WW2 than the throbbing
drone of an approaching flying bomb
and the heart-stopping and eerie
silence between the engine cutting out
and the impact seconds later. That
distinctive sound from the Argus pulse
jet (often likened to the sound of a
motorcycle) earned the V-1 the
widespread nickname of *Buzz Bomb*
with the public soon after the first fell
in southern England on 13 June 1944,
and the equally popular title of
Doodlebug originated with the Kiwi
pilots of 486 (NZ) Squadron RAF,
who named it after a clumsy species of
insect native to their homeland.
Earlier, it had been given the
codename Peenemunde 20 (after its
estimated 20ft span) by Constance
Babington-Smith of the Allied Central
Interpretation Unit when first
photographed at the German secret
research establishment, and of course
P-Plane simply meant 'Pilotless
Plane'. The V-1 was also designated
the Fieseler Fi 103 in Germany, and
roughly 21,000 were launched against
targets in Britain and Belgium, the last
being shot down over Kent on 29
March 1945. Much less well-known,
even in Germany, was the effort put
into developing a manned version,
known in its intended operational
form as the Reichenberg IV (the Mks
I to III being trainers). Theoretically,
the pilot would bale out after aiming
it at the target, but in reality it was a
suicide mission little removed from
the Japanese kamikazes, and it was

controversy and in-fighting over this
which ensured the **Reichenberg** was
never used operationally, although
something like 175 machines were
completed.
   As early as October 1944 the
USAAF began testing a remarkably
faithful copy of the V-1, built by
Republic Aviation as the JB-2 and
nicknamed *Thunderbug* in keeping
with the company's 'Thundercraft'
names. Post-war, the project was
taken up by the US Navy as the Loon,
and altogether some 1,200 JB-2s were
built in the US.

Reich's Lighter
One of the more innovative bombers
of WW2 was the Heinkel He 177 Greif
(**Griffon**) with its four engines coupled
in pairs, each pair driving a single
large prop, which gave the aircraft the
appearance of a large underpowered
twin. More significantly, the
powerplant arrangement led to
overheating problems and a
reputation for inflight fires, reflected
in such epithets as the *Reich's Lighter*,
Luftwaffe's Lighter and the usual
Flaming Coffin. During the bloody
battle for Stalingrad in the winter of
1942/43 a number of He 177s were
fitted with a 50mm anti-tank gun in
the nose gondola by a forward
maintenance unit; later the He 177A-
3/R5 sub-variant appeared with a
similarly mounted 75mm weapon, and
though just five were built the mark
was known as the *Stalingrad Type* in
recognition of the earlier field
modification.

Reliant Robin
With a new tricycle-wheeled
undercarriage in place of the skids
used on earlier models, the Westland
Lynx AH9 helicopter is known within
the British Army Air Corps as the

Reliant Robin, after the much-derided small car. In contrast, the fast and racy Lynx was sometimes known in its early days as the *E-Type Helicopter*, after the classic E-Type Jaguar sports car.

Reporter
Designated in the 1930 'F-for-Photo-Recce' sequence, the Northrop F-15A **Reporter** was a relatively minor development of the famous P-61 Black Widow night-fighter, with a new two-seat cockpit and bubble canopy. The unarmed F-15A was actually based on the airframe of the XP-61E day-fighter project (two prototypes only) and a total of just 36 were built in 1946, becoming RF-61Cs two years later.

Flying Restaurant
According to Viktor Belenko, who defected to Japan with his MiG-25 (NATO – Foxbat) in 1976, the aircraft needed half a ton of alcohol (for its electronic and braking systems) on a typical flight, which naturally made MiG-25 bases a very popular posting for Soviet Air Force personnel. Legend has it that air force wives protesting to an unsympathetic design bureau in Moscow were informed that the next MiG would be fuelled with pure cognac.

Rhapsody in Glue
Resembling an under-developed Anson and derived from the pre-war Cessna T-50, the USAAF's wooden UC-78, AT-8 and AT-17 Bobcat trainers and transports must have been the most parodied aircraft in their class. The UC-78 staff transport became the *Useless 78* or *Brasshat*, whilst the line as a whole was known to servicemen as the *Double-Breasted Cub* (after the Piper Cub), *Bamboo*

Bomber, Boxkite and (with apologies to George Gershwin) the immortal *Rhapsody in Glue*. 550 AT-17As were to have been supplied to Canada as the Crane, but deliveries stopped at 182 when America entered the war.

Rocket (i)
Before the Product Liability Laws forced its suspension in late 1987, Cessna and licensee Reims Aviation built almost 36,000 Model 172 light aircraft between them, making it by far the world's most popular private-owner aircraft and a strong contender for the outright title of the world's most prolific aircraft. Skyhawk is often mistaken for an overall type name for the 172, but like many Cessna names actually applied strictly to the de-luxe models, and is parodied in the colloquial form of *Chickenhawk* applied to the USAF's T-41 Mescalero trainer version. Other mainstream variants built by Cessna include the higher-powered Model R172K Hawk XP, retractable-gear Model 172RG Cutlass, and Continental-powered Model 175 Skylark. Dart I was the title of a fairly straightforward conversion by Cessna's Wichita neighbours Doyn Aircraft, involving the fitting of a more powerful Lycoming engine and a constant-speed propeller. In addition to large quantities of standard American models, Reims Aviation also built 590 of their own higher-powered **Rocket** (strictly the Reims **Rocket**) version from 1967.

Rocket (ii)
Not to be confused with the Reims **Rocket** (see previous entry) is the long series of conversions to Cessna twins by the former Riley Aircraft Manufacturing Inc. of Carlsbad, California. The original 1962 Riley

Rocket was a Cessna 310 with new 290hp Lycomings (the Turbo-Rocket had Rajay turbo-super-chargers), three-blade props, improved sound proofing, and a host of minor aerodynamic refinements. It was followed in the seventies by the Rocket 340 and Rocket 414, Cessna 340s and 414s respectively with new 340hp and 400hp turbo-super-charged Lycomings. Riley also matched the classic Cessna 421C Golden Eagle airframe to the ubiquitous P&WC PT6A-135 turboprop in 1979 to produce the Turbine Eagle 421; Advanced Aircraft Corp. renamed it Regent 1500 when they took over part of the bankrupt Riley Aircraft operation in 1983, but the original title survived for an almost identical conversion offered by the new Riley International. The new Riley outfit also developed the Rocket P-210, a conversion of the single-engined Cessna P-210 Pressurised Centurion, with a new engine intercooler system, avionics refit, and luxury interior.

Rocking Chair
Rocking Chair and *Gondola* were both common names for the WW1 Austrian Aviatik BIII recce biplane, notoriously sluggish on the controls and thus prone to swinging around in windy weather.

Rook
Grach (*Rook*) is the semi-official name favoured by groundcrews in the Russian Air Force for the Sukhoi Su-25 close-support aircraft (NATO – Frogfoot) and derives from the split airbrakes at its wingtips. Pilots reportedly prefer *Cherabushka* or *Little Critter* for the Su-25, which performs a broadly similar role to America's Fairchild A-10.

Rota
RAF name for the dozen Avro-built Cierva C30A Autogiros delivered from August 1934. Rota II was the title of the improved C40 built at Hanworth by the British Aircraft Manufacturing Co. (BA), differing by having an autodynamic rotor-head permitting jump-starts instead of the more usual short ground-run. The C30 was also licence-built in Germany by Focke-Wulf as the Heuschrecke (**Grasshopper**).

Royal Gull
American marketing name (later shortened to just Gull) used by Kearney & Trecker for the Piaggio P136 twin-engined light amphibian, in low-key production in Italy from 1948 to 1967. The P136's tail, gull wings and pusher engine installation were also married to a deeper fuselage to form the P166 landplane, known in P166B form only as the Portofino, and to the SAAF as the P166S Albatross.

Sabre Slayer
Developed alongside the short-lived Fo 139 Midge of 1954, the Folland Fo 141 Gnat was more than just an RAF advanced trainer (Fo 144) and Red Arrows mount from 1965 to 1979. The original *Pocket Fighter* was exported to Finland (twelve), Yugoslavia (two for evaluation) and India (40), where HAL also built 213 under licence. In the space of twenty days during the 1965 Indo-Pakistan conflict, the tiny transonic Gnat shot down at least six examples of the battle-proven North American F-86, earning the almost reverentially applied title of *Sabre Slayer* throughout India, not to mention commemoration by dozens of 'Gnat Brand' products. HAL also developed the Gnat Mk II or Ajeet (variously translated as 'Invincible',

'Unconquerable' or 'Unconquered')
with a wet wing and improved
controls, which served the IAF in
relatively modest numbers until 1991.
There was also a two-seat Ajeet
trainer, differing from the RAF's
Gnat T1 by having individual cockpit
hoods.

Safari
Production of the Soviet Antonov An-
28 utility transport (NATO – Cash)
began in Poland in 1978, and in July
1993 PZL Mielec flew a new export-
oriented version, initially known as
the An-28PT **Safari**, with new
Western avionics, P&WC PT6A-65B
turboprops, and five-blade Hartzell
props. A separate An-28
development, retaining the original
Rzeszow TVD-10B engines, is the An-
28B-1R Bryza (Breeze) coastal patrol/
SAR aircraft for the Polish Navy with
indigenous SRN-441XA search radar
in a ventral radome.

Safety Pin
Reference to both the maker's name
and the twin-boom layout of the Arpin
A-1, a one-off (G-AFGB) pusher
lightplane flown in Britain in 1938.

Salon
Soviet/Eastern Bloc suffix for VIP or
luxury models of a standard type, for
example variants of the Avia 14
(Czech-built Ilyushin I1-14, NATO –
Crate), the Mil Mi-8 helicopter
(NATO – Hip) and more recently the
Polish PZL Swidnik Sokol helicopter.

Salvage
Pun on the name of the US Navy's
post-war era North American AJ
Savage, a carrier-borne bomber
powered by twin P&W R-2800 Double
Wasp radials and a single Allison J33
turbojet in the tail.

Schwalbe
In contrast to its British counterpart
the **Meteor** (cf *Meatbox*), relatively
little effort was apparently expended
on naming the Messerschmitt Me 262
(usually regarded as the world's first
true jet fighter), and even the oft-
quoted title of **Schwalbe** (Swallow)
was never formally ratified.
Sturmvogel (Stormbird) was more
formal however, and was used to
differentiate the Me 262A-2a fighter-
bomber from the pure interceptor
models, following Hitler's famous
edict over the role of the Me 262.
More generally, pilots usually referred
to the Me 262 simply as the *Turbo*.

Scipio
Imperial Airways class name for its
trio of 1931 Short S17 Kents, four-
engined biplane flying boats noted for
their ornate cabin furnishings; flagship
of the fleet was G-ABFA.

Screwball
Occasional term for the Royal Navy's
Blackburn Skua two-seat fighter-dive-
bomber. Like many Blackburn types,
the Skua was not among the more
popular aircraft of its day, but is
historically important as the Royal
Navy's first production monoplane,
and as the type which scored the first
kill by a British-built aircraft of WW2,
when an 803 Squadron aircraft
downed a Dornier Do 18 off Norway
on 25 September 1939. A major
variant of the Skua was the Roc, built
by Boulton Paul, and featuring a
power-operated, four-gun turret by
the same company.

Sea Dragon
Sikorsky's multi-faceted S-61
helicopter family first appeared in
March 1959 as the US Navy's HSS-2
Sea Dragon, but in 1962 the title was

changed to the more familiar SH-3 Sea King. The HSS-2 tag was for 'funding purposes', suggesting an improved Seabat (cf *Hiss*), and in fact the larger machine was briefly nicknamed *Hiss Two*; curiously, the HSS-2 designation is still used in Japan, where the Sea King also has the local name of Chidori (Plover). During Vietnam operations the SH-3A and HH-3A also picked up the immortal *Big Mother,* not in recongnition of their search-and-rescue role, but after a harassed carrier captain allegedly screamed at someone to 'get those (expletive deleted) Big Mothers off my deck!'

The first true USAF model was the CH-3C *Big Charlie* which introduced the rear-loading ramp, but of course the best known air force model was the HH-3E combat-rescue helicopter, dubbed *Jolly Green Giant* in Vietnam after the brand of sweetcorn; although obviously originating as a nickname, the title was later given official recognition. Outwardly similar in appearance (except for the colour scheme and nose radome) was the US Coast Guard's HH-3F, which has the formal but rarely used name of Pelican; the Italian variant of the HH-3F with LORAN-C aerials protruding under the nose like tusks is known as the *Walrus*. Malaysia also acquired no less than 40 land-based transport models, which they named Nuri (Yellow Bird), but these are 'boat-tailed' S-61A-4s.

The stretched commercial models were the S-61N (for 'Nautical', i.e. for overwater operations) and S-61L (Land-based). Both were boat-tailed, and the S-61L was distinguished by its lack of sponsons; however, a limited edition, stripped-down crane version of the S-61N, the Payloader, also lacked sponsons. Bristow Helicopter

S-61s operating under contract in the Falklands (post-war) were dubbed *Erics* by the squaddies, after the 'Crafty Cockney' darts player. Sikorsky closed the civil line somewhat prematurely with the 136th example in 1980, but the attempt by licensee Agusta to fill the gap with their shortened, increased-fuel AS-61N-1 Silver proved disappointing; Malaysia was the only customer, taking two in VIP configuration.

Westland's interest in the S-61 family predates their Sea King production with the almost forgotten 23-seat civil Wiltshire proposal, circa 1959. In 1973 the British licensee first flew their export Commando, a land-based (though still boat-tailed) transport derivative, followed in 1979 by the first of 41 broadly similar Sea King HC4s for the Royal Navy; these inherited the popular nickname of *Jungly* from the Wessex HU5s they supplanted as the Royal Marines' transport helicopters.

Sea Guard
Rarely used name for the US Coast Guard's Sikorsky HH-52A helicopter. Resembling a shortened, single-engined Sea King, the HH-52A was a derivative of the commercial S-62, and entered service with the Coast Guard in January 1963.

Seamew
Royal Navy name for the WW2 Curtiss SO3C scout monoplane, operable with either floats or fixed, wheeled undercarriage. The name was later taken up by the US Navy in place of the unofficial Curtiss name of Seagull.

Self-Jamming Bomber
Initiated under Nixon, cancelled by Carter, and resurrected by Reagan,

the Rockwell B-1 swing-wing strategic bomber has been one of the USAF's more controversial aircraft of recent times. Critics said the downgraded B-1B production model was altogether too late and too expensive, engine problems made it the only major US type to miss the 1991 Gulf War, and glitches in the ambitious AN/ALQ-161 defensive avionics suite were personified in the withering nickname of the *Self-Jamming Bomber*; a less common nickname is Bone, from B-One. Controversy degenerated into farce in 1989, when the Air Force publicly rejected a plan by a Congressman to have it christened the Excalibur, on the not unreasonable grounds that the name was already too well-established in the States as a best selling brand of condom. The B-1B was not finally named Lancer until 1990. Production of 100 B-1Bs ended in 1988, and despite its problems the type has an important place in the USAF inventory, filling a gap between the ageing B-52 and the severely-curtailed B-2 Spirit.

Seneca
US Army name for the ten Cessna CH-1B light helicopters it evaluated at Fort Rucker, Alabama from March 1957 under the designation YH-41. Cessna's flirtation with helicopters began in 1952 with the acquisition of the Seibel Helicopter Company, and the all-new CH-1 followed in 1954, bearing a distinct family resemblance to Cessna lightplanes with its cabin arrangements and nose-mounted piston engine. Despite an impressive performance only around 29 were built including the ten YH-41s, four for Ecuador and five for Iran, the export machines being variants of the

definitive civil CH-1C Skyhook. Production ended in 1962, after the company decided that the civil helicopter market was not then mature enough to warrant further investment.

Sequoya
Having taken over the classy Ted Smith Aerostar business twin in 1978, Piper elected to continue its tradition of Indian titles by naming the new Model 602P the **Sequoya**; the name proved unpopular with Aerostar owners however, and was dropped following protests from Sequoia Aircraft Corp. (marketing their Model 300 for home-building) who had taken their better known spelling variation from the sequoia redwood tree. Ted Smith Associates named their more powerful and slightly larger Model 700 prototype of 1972 the Superstar, and the title was later taken up by Machen Inc. of Washington for a series of performance upgrade packages featuring new turbochargers, improved pressurisation, and changes to the fuel system. In 1984, Machen announced its Laser proposal, which would have seen the usual piston engines of the Model 601P replaced by Garrett TFE76 turbofans in nacelles on the rear fuselage; sadly the project came to nothing, as did the 1991 Star Jet proposal by Machen associates Aerostar Aircraft (who bought the rights from Piper) with underslung Williams/Rolls-Royce FJ44 turbofans.

Seraph
Scout was apparently the first name considered for the Westland (Saro) P531 light helicopter, and of course survived for the Army Air Corps

114

(AAC) version ordered in 1960. The
Royal Navy, meanwhile, toyed with
the names **Seraph**, Sprite, Sea Sprite
and Sea Scout before finally settling
for Wasp.

Seven Day Bus
One of the many tales concerning the
remarkable Noel Pemberton-Billing is
of how on the day the Great War
broke out (4 August 1914) he called
together the staff at his 'Supermarine'
works in Southampton, and
announced that no one would be going
home until they had designed and
built a new fighting scout. The
resulting PB9 (one only) went down in
aviation history as the *Seven Day Bus*,
and whilst more recent research
suggests it was not built entirely from
scratch in that time, that should not
detract from either the achievement or
the point it made.

Seven Seas
Beautifully apt name used by Douglas
for their final propliner model, the
DC-7C, reflecting not only its alpha-
numeric designation but also its status
as one of the first truly long-range
airliners; in recognition of the name,
KLM's DC-7C simulator at Schipol
was colloquially known as the Dry
Sea. Pan Am promoted its DC-7Bs
and 7Cs for a while as Super 7s (cf
Carvair 7, under **North Star**).

Sewing Machine
First flown as the U-2 on 7 January
1928, the Polikarpov Po-2 (NATO –
Mule) was in some respects the Soviet
counterpart to the Avro 504 or Curtiss
Jenny although in terms of production
and longevity it leaves them both
standing, with an estimated 40,000
built before production finally ended
in Poland in 1953. The most popular
nicknames were *Sewing Machine*

(after the sound of its five-cyclinder
M-11 radial engine), *Grasshopper* and
Corn Cutter. A firefighting version
was known more formally as the
Lesnik (Forest Guard).

Shack
They used to say that the Avro
Shackleton was 'as ugly as a box of
frogs' and dubbed it the *Growler*
because of the din from its Griffon
engines and contra-rotating props (the
MR3 series 3 also had two Viper jets),
but as the last of the piston heavies its
final retirement in June 1991 was
widely mourned. The *Shack* was an
outgrowth of the Lincoln bomber
(initially designated Lincoln 3) and
served the RAF faithfully in its
original maritime role until the arrival
of the Nimrod in 1969. Maritime
aircraft fitted with the unusually
shaped dorsal aerial were also known
as *Flying Spark Plugs*, but it was in the
final Airborne Early Warning AEW 2
form with 8 Squadron RAF that the
type picked up such affectionate
monickers as the *Old Grey Lady*,
Shacklebomber (denoting the
Lancaster/Lincoln ancestry), the
Contra-rotating Nissen Hut, and the
Bear Hunter, after the Tupolev Tu-
95s & 142s (NATO – Bear) they
encountered off the coast of Scotland.
The most original though must be the
Magic Roundabout, with individual
aircraft named 'Florence', 'Mr Rusty'
etc., after characters in the children's
TV show.

Shagbat
One of the more individual aircraft of
WW2, the Supermarine Walrus
biplane amphibian had the suitably
distinctive and ubiquitous nickname of
Shagbat, supposedly because it was
shaggy like a walrus and flew like a
bat. Lesser titles were the *Flying Gas*

Ring from the glow of the uncowled pusher engine at night, *Steam Pigeon* from the effects of water spray hitting the hot engine, and *Duck* from its amphibious capability. Though it is perhaps best remembered for its wartime air-sea rescue exploits, the Walrus first flew in 1933 as a shipborne, catapult-launched spotter for the RAAF and was named Seagull V after Supermarine's earlier amphibians; Australia kept the Seagull V title, and the RAF/RN name of Walrus did not appear until August 1935.

Shamsher
Indian Air Force name (pronounced 'shamsheer' and meaning 'Assault Sword') for the Anglo-French Sepecat Jaguar strike aircraft. Britain supplied 40 complete aircraft plus 45 in component form, and a further 31 have been built by HAL at Bangalore. The export version of the Jaguar is known to its makers as the Jaguar International, and other customers apart from India included Ecuador (twelve), Nigeria (eighteen) and Oman (24).

Shark
With a slimmer fuselage than its stablemate the **Whale**, the streamlined LFG ('Roland') DII German scout of WW1 was given the complementary nickname of *Haifisch* or **Shark**.

Flying Shoe
Widely used nickname, along with *Flying Clog*, for the Blohm und Voss BV 138 twin-boom flying boat, after the unusual shape of the short fuselage. The BV 138 MS minesweeping variant was known as the *Mausi-Flugzeuge* or *Mouse-Catcher* aircraft.

Shorthorn
Yet again, C.G. Grey, Editor of *Aeroplane*, seems to have been instrumental in two of WW1's most famous and widespread nicknames. The RFC had already given the nickname *Mechanical Cow* to the French-designed Maurice Farman 7 biplanes they began receiving in the summer of 1912, which were notable for the way their large landing skids extended forwards to carry the front elevator. Late the following year the Maurice Farman 11 variant appeared in England with no front elevator and consequently much shorter skids; Grey's suggestion that this must be the ***Shorthorn*** breed of the aforementioned *Cow* was quickly taken up in the RFC, and the earlier MF 7 was retrospectively dubbed *Longhorn* to differentiate between the pair. Almost as common was *Rumpety*, which covered both types and is probably onomatopoeic, and *Flying Incinerator* appears to have had more to do with the early aviators' natural fear of inflight fires than any particular characteristic of the Farmans. *Chicken Coop* was also used for the MF 7, and in *Flying Fury* James McCudden notes the sarcastic term *Longhorn Bullet* (cf **Bullet**), but despite their archaic appearance and their reputation as an 'aerial joke' (as they were once described) both played an important part for the RFC and RNAS in the early years of WW1, particularly as trainers.

Shrew
The most famous quote on aircraft names surely belongs to the legendary R.J. Mitchell who, upon hearing that his new Type 300 fighter was to be called Spitfire, reputedly described it as 'just the sort of bloody silly name they *would* give it.' **Shrew** (in the

aggressive, Shakespearean sense) was also considered, but in fact the Spitfire name had already been loosely associated with the earlier Supermarine Type 224 fixed-gear, gull-winged prototype of 1934; as a result the Type 300 (i.e. *the* Spitfire) was briefly known as the Spitfire II in 1936, the year it first flew. The name quickly became such a legend that it seems incomprehensible that consideration was given to renaming the Spitfire F21 and unbuilt F23 the Victor and Valiant respectively, presumably to provide some alliterative connection with parent company Vickers. Victor was also used briefly in 1943 for the Supermarine Spiteful, but whilst it and its naval counterpart the Seafang were clearly offshoots of the Spitfire line, their new laminar-flow wings made them virtually new aircraft rather than genuine variants. The Seafire was of course the navalised deck-landing variant of the Spitfire, originally known by the obvious and simple title Sea Spitfire.

As for nicknames, *Spit* and the early-war variation *Spitter* require no comment, and the *Narvik Nightmare* was the unhappy floatplane project of 1940, intended for operation from Norwegian fjords; the scheme was gladly dropped with the end of the Norwegian campaign, although a more successful 'booted' Spitfire was devised by Folland Aircraft in 1942. Two more abortive projects were the Speed Spitfire (N17) which flew in 1939 but never made its intended attempt on the world air-speed record, and the Aerolite Spitfire aimed at preserving strategic materials by the use of bonded plastics, but which progressed no further than a Spitfire V-type fuselage static-tested at Farnborough.

Bomfires were simply Spitfires operating as fighter-bombers with bombs or rocket projectiles, and as a publicity stunt the fighter was also employed in 1944 to carry 'Mod XXX Depth Charges' – barrels of Henty & Constable beer to boost the morale of the troops in Normandy.

The Spitfire story is of course inextricably linked with that of the Merlin engine (later marks used Griffons) but therein also lies one of the big fallacies of British nomenclature: like all other Rolls-Royce piston aero-engines the Merlin took its name from a carrion bird and not, as so often stated, from King Arthur's magician.

Shturmovik

Just as the German term *Stuka* became synonymous with the Ju 87, so this famous Russian title (translating roughly as 'Assaulter') has become indissolubly linked with the Ilyushin Il-2 (NATO – Bark), although in its homeland it was also associated with other types, notably the rival Sukhoi Su-2. One of the great symbols of Soviet resistance in WW2, the Il-2 was supposedly known to the troops it supported as the *Flying Tank*, *Flying Infantryman* or *Hunchback*, and to their opposite numbers as the *Black Death*; more dependable is the German term *Iron Gustav* (after the Bf 109G) quoted in Hans Ulrich Rudel's *Stuka Pilot*. To those who actually flew it, the Il-2 was usually just the Ilyusha, as commemorated in a popular song of the times, 'Me and my Ilyusha'. Combined output of the Il-2 and improved Il-10 (NATO – Beast) has been estimated as high as 42,000, including post-war production in Czechoslovakia.

Silly Grin Aeroplane, The

Name associated with the Shorts Tucano turboprop trainer around the time it first entered service with the Central Flying School (CFS) in 1988, noting the delights to be had from its excellent handling and its excess power. When the Brazilian Embraer EMB-312 Tucano was first selected to be licence-built for the RAF in 1985, it was widely expected that it would be renamed in due course, or at the very least anglicised to Toucan.

Silver Arrow

Title reportedly adopted by Aeroflot for the Tupolev Tu-104 (NATO – Camel) twin-jet airliner in 1956, although little has been heard of the name since. Supposedly unknown in the West, the Tu-104 caused something of a sensation when it made a surprise visit to London Heathrow on 22 March 1956, particularly as the DH Comet 1 had been grounded in 1954 and the Boeing 707 was not to enter service for another two-and-a-half years. It has been suggested that Western observers failed to note the type when it took part in the previous year's Tushino flypast, but in time it emerged that the 'Tu-104G development aircraft' thought to have been involved was actually just a civil-registered Tu-16 (NATO-Badger), the familiar bomber on which the real Tu-104 airliner was based. The true nature of the 'Tu-104G' (alias Tu-16G) also explained the *Little Red Riding Hood* nickname, which had baffled the West for something like twenty years.

Silver Hawk

Occasional RFC nickname for the popular French-built Nieuport 17 scout, flown by such aces as Albert Ball, Edward 'Mick' Mannock, and Billy Bishop.

Silver Sow

Whilst the USAF's Boeing KC-135 Stratotanker and much less prolific C-135 Stratolifter may appear outwardly to be relatively straightforward variants of the famous 707 airliner (cf **Jet Stratoliner**), there are subtle but important differences in materials and structure, as reflected in the fact that they also bear the 'missing' Boeing model number of 717. The tanker/transports were also the first to enter service in June 1957, and both they and the 707 have their origins in the famous yellow and brown Model 367-80 'Dash-80' prototype (N70700), so designated for reasons of industrial camouflage to suggest a revised Model 367/C-97.

In service, the KC-135 tanker has been nicknamed *Silver Sow* from the way it 'suckles' receiving aircraft, along with *Stratobladder*, **Lead Sled**, and *GLOB* from 'Ground-Loving Old Bastard'. As well as 45 C-135A Stratolifter transports built as such, there were also three *C-135 Falsies* produced by removing refuelling gear from KC-135s. The KC-135 has of course been adapted to a multitude of other roles, and some variants have been so festooned with aerials, antennae, radomes, 'Hog Noses', 'Platypus Noses' and sundry other classified lumps and bumps that the ordinary 'vanilla' tanker acquired the title *Plain Jane* to separate it from its more exotic siblings. One of the best known adaptations must be the various EC-135 Airborne Command Posts (notably in the Cover All and Looking Glass programmes) flown around the clock to ensure that a degree of military command would survive in the event of a surprise nuclear attack; with all their chilling 'Dr Strangelove' connotations, these have been widely referred to as

Doomsday Planes. Some of the more severely doctored KC-135s come within the extensive range of RC-135 electronic reconnaissance aircraft (the Cobra Ball, Combat Sent, Rivet Joint programmes, etc.), and although *Ferret* really belongs to the mission of 'sniffing out' radar and communications signals, the name is also used informally for the actual aircraft used. One of the first reconnaissance models, the KC-135R (not to be confused with the later tanker upgrade to CFM-56 turbofans) was nicknamed the *Porcupine* on account of what in retrospect seems a modest collection of aerials along its back. Another mark-specific nickname is *Piccolo Tube* for a particular aircraft (55-3135) in the NKC-135A research series with a long cylindrical fuselage spattered almost shotgun-style with windows of various sizes. In contrast, the C-135FR tankers of the French Armée de l'Air (the only foreign operator) have been nicknamed *Submarine* because of their lack of windows, along with *Monster* and **Whale**.

Silver Swallow
Occasional and unofficial name within the North Vietnamese Air Force for the MiG-17 (NATO – Fresco) fighter, known in codified form to US forces as the Red Bandit.

Six Shooter
Originating as the formal title of the programme to fit an M61 Vulcan 20mm cannon to the USAF's Convair F-106 Delta Dart interceptor in the early seventies, **Six Shooter** inevitably became a colloquial term for actual aircraft thus modified.

Sky Cleaner
Also described more realistically as a weather modification aircraft, the Antonov An-30M **Sky Cleaner** is a minor variant of the standard An-30 (NATO – Clank) photo-survey aircraft, itself a derivative of the prolific An-24, An-26 and An-32 family of twin-turboprop transports. The weather can supposedly be 'modified' by seeding clouds with granulated carbon dioxide, producing rainfall over farmland or forest fires, or diverting it from urban areas.

Skyliner
Class name suggested by the Aeronautical Engineers Association in 1953 for BOAC's de Havilland DH 106 Comet jet airliners; the individual name of 'Sky Courier' was suggested for the flagship of the fleet, but of course neither name was taken up.

Flying Skyscraper
Media term for the famous Bristol 167 Brabazon airliner. First flown on 14 September 1949, the 230ft-span *Brab* remains Britain's biggest ever aircraft, and one of its biggest white elephants.

Skystar
Name adopted by Ansett Airlines of Australia (now Ansett Australia) for the Airbus A320 twin-jet airliner, the first of which was handed over to the carrier on 18 November 1988.

SLUF
Possibly the best known of all acronym type nicknames is the *SLUF* title applied to the popular Vought (LTV) A-7 Corsair II attack bomber, supposedly an abbreviation for 'Short Little (or occasionally Low) Ugly Fellow'. Proponents of the F/A-18 **Hornet** (which replaced it in the Navy) changed the first word to Slow, and *Supersluf* was sometimes used for the Air Force A-7D and Navy A-7E

models, which introduced the Allison
TF41 engine (licence-built Rolls-
Royce Spey) in place of the earlier
P&W TF30. Other versions must be
left to the reader's imagination.

Developed initially for the Navy,
the Corsair II was of course named
after the WW2 F4U (which itself could
have been the 'II', after the O2U,
O3U and SU Corsair biplanes) and
the strong naval connotations may
have had a bearing on the fact that the
name was not taken up for the Air
Force A-7D and National Guard A-
7K, even informally. In south-east
Asia the A-7D proved particularly
effective in 'Sandy' Rescap (Rescue
Combat Air Patrol) operations, and as
with the *Spad* it supplanted, the *Sandy*
title became informally associated
with the A-7D itself. Proposals to
upgrade the A-7D and K to a
supersonic capability carried the
successive titles A-7X, Fast LANA
Strikefighter and A-7 Plus (LANA
was the Vought Low-Altitude Night-
Attack package already fitted to 83
National Guard aircraft). Two
airframes were eventually modified as
YA-7Fs with a stretched fuselage,
extended fin, and a 23,000lb thrust
P&W F100 turbofan giving almost 50
per cent more power than the subsonic
models; the first flew in November
1989, but no further conversions were
undertaken.

Snake
Bell's early thoughts on a dedicated
two-seat gunship helicopter using the
dynamic components of the UH-1
Huey found expression in the D-255
Iroquois Warrior project, which
appeared in full-scale mock-up form in
1962. By the time the definitive Model
209 flew in 1965 the unofficial name of
Cobra had become associated with
armed *Hueys* in Vietnam, and so the

company chose to combine the two
names into the evocative title of
HueyCobra; the Army grudgingly
went along with plain **Cobra** for what
was by then the AH-1G, although in
the field it is often simply the *Snake*.

SeaCobra was the obvious title for
the AH-1J and AH-1T tailored to
Marine Corps requirements with
PT6T Twin-Pac powerplants, and the
SuperCobra is the Marines' greatly
enhanced AH-1W; a further
development of this 'Whisky' model is
the AH-1(4B)W Viper (CobraShark
was also considered) with a
completely new four-blade,
bearingless rotor, flown in prototype
form in January 1989. The slightly
stretched Model 309 KingCobra of
1971 was optimised more for
European anti-tank operations with
TOW missiles, but it did not progress
beyond the prototype stage. Cobra
Venom is the name of the variant
competing in 1994 for a British Army
order, and as part of the sales pitch,
Bell PR people have used the term
Flying Anvil to describe its strength.

Snargasher
The Reid & Sigrist RS1 twin-engined
trainer (G-AEOD) received its
delightful nickname as a 'family joke'
during construction in 1938; only one
was built, but in the wartime RAF the
name was also sometimes used for
training aircraft and trainee crews in
general, supposedly as a corruption of
'Tarmac Smasher'.

Snub Nose
The Soviet Union supplied in excess of
200 Polikarpov I-15 Chaika (Seagull)
fighter biplanes to the Republicans in
the Spanish Civil War, where they
were given the derogatory title *Chato*
(*Snub Nose*) by the opposing
Nationalists. As a result of the

commonly-held belief that the Soviet Union was incapable of anything more than copying Western types, the Spanish I-15s were frequently referred to as the 'Curtiss'.

Space Ship
Marketing title used by American Eurocopter to emphasis the roominess of the MBB/Kawasaki BK 117, the eight- to twelve-seat German/Japanese helicopter first flown in June 1979.

Spad
Supposedly designed overnight in a Washington hotel room and first flown in March 1945 as the XBT2D Dauntless II, the Douglas AD Skyraider (as it was renamed in February 1946) single-seat attack bomber served the US Navy for so long that it became fondly known as the *Spad* after the French scouts used by the Air Service in WW1. The legendary endurance and load-carrying abilities of the piston-powered AD also earned it such accolades as the *Big Gun*, *Old Faithful* and *Old Miscellaneous* (to prove the point, kitchen sinks were dropped on both Korea and Vietnam), and the stretched and widened AD-5 with the 'Blue Room' compartment behind the cockpit was sometimes known as the *Fat Face* version. Specialised models were the Early Warning AD-3W to AD-5W *Guppies* (named after the belly radome) and AD-1Q to AD-5Q *Q-Bird*s.
   In Vietnam the *Spad* was further immortalised by its 'Sandy' Rescap (Rescue Combat Air Patrol) missions with the USAF, and in fact the *Sandy* name rubbed off onto the aircraft itself to some degree. The air force also knew the supremely versatile (strike to twelve-seat transport) A-1E

model as the *Flying Dumptruck*, whilst the South Vietnamese called it the *Crazy Water Buffalo*, but in the end perhaps the most fitting of all colloquial terms for the venerable old Skyraider was simply the phonetic version of its pre-1962 designation of AD – *Able Dog*.

Sparklet
Pemberton-Billing's experimental PB 23E fighter of 1915 acquired its nickname because its tiny fuselage nacelle was thought to resemble a Sparklet soda-siphon bulb. An alternative nickname was the hopeful *Push-Proj* (from 'pusher-projectile') which was occasionally spelt Push-Prodge after test pilot C.B. Prodger. A much modified version went into limited production for the RNAS as the PB 25, but it is not thought that any of the twenty built were ever used operationally.

Sparrow (i)
Informal name of the bizarre A.D. Scout, designed as a single-seat, anti-Zeppelin fighter by Harris Booth of the Admiralty Air Department, and intended to be armed with a Davis two-pounder recoil-less gun on the cockpit floor. The pusher-engined *Sparrow* featured a nacelle attached to the underside of the upper wing, an inordinately large interplane gap, an extremely narrow-track undercarriage, and a tailplane of no less than 21ft span. A curious characteristic of the Davis gun was that on firing it occasionally produced a rearwards discharge of Epsom salts, although to any pilot reaching the point of combat in such a strange machine, any side-effects would probably already be superfluous. Thankfully only two were definitely completed (by Blackburn), although

the status of a further two from Hewlett & Blondeau is uncertain.

Sparrow (ii)
Rhyming slang for Handley Page HP 54 Harrow bombers converted for use as transports with 1680 Flight, Fighter Command/271 Squadron RAF in 1940. Five Harrows were also used in Operation Mutton to develop the Long Aerial Mines (LAM) intended for operational use with specially converted Havoc night-fighters (cf **Boston**). The LAMs were dropped in quantity by parachute, trailing a 2,000ft length of wire which would theoretically entangle the enemy bombers and bring the charges down upon them. A handful of kills were reported, but the poor old Harrows themselves took the majority of hits, and aircraft returning to Middle Wallop with their fabric covering stripped away as a result were known as *Steam Chickens*.

Spartan
C-27 **Spartan** is the USAF designation for the Alenia G222 twin-turboprop STOL transport, built in Italy but with a 60 per cent US content; delivery of an initial ten C-27As began in 1991. The G222 was originally conceived by Fiat (later Aeritalia) as a V/STOL type in 1963, with the short-lived name Cervino.

Spider
Nickname of the experimental Vickers FB12 pusher scout of 1916, after its long and stalky undercarriage.

Spider Crab
Project-phase name for the de Havilland DH 100 Vampire jet fighter. The twin-boom layout of the Vampire led to the occasional nickname of *Flying Wheelbarrow*, but

much more prevalent in the RAF was the derisory *Kiddie Kar*, after its exceptionally low sit on the ground and its basic first-generation jet technology (not that that stopped Meteor pilots joining in too). In Mexico, with the plump and rounded nacelle painted a pretty shade of green, pilots christened it the *Aguacate* (*Avocado*). The biggest foreign user though was France, where SNCASE assembled 250 Goblin-engined Vampire FB5s (67 of them from British parts), followed by 247 of the locally developed SE-535 Mistral variant with French-built RR Nene engines and Martin Baker ejection seats.

In February 1968, at a Motor Show-style launch in London, complete with a mock-up adorned by four scantily-clad ladies, Jet Craft of Las Vegas unveiled plans to turn 36 surplus airframes into two-seat Mystery Jet 1 civil trainers and six-seat Mystery Jet 2 executive aircraft. The actual conversion work was sub-contracted to Marshalls of Cambridge, but the project collapsed with the bankruptcy and subsequent jailing of Jet Craft's founder. The six-seater project re-emerged in 1979 as the Executive Mk 1 (P&W JT15D-5 turbofan) and again a decade later when a version with a PT6A-65AR pusher turboprop, the Jet Cruiser, was also envisaged; by this time the intention was to use newly-manufactured Vampire components, but yet again the project foundered.

Spinning Incinerator
Later to become famous as the mount of Rees and Hawker, the RFC's Airco DH 2 pusher scout of 1915 had an early reputation for spinning crashes, often attributed to control over-sensitivity and engine disintegration.

2

SPINNING JENNY

Spinning Jenny
RNAS nickname for the generally unloved and otherwise unnamed Sopwith Two-Seat Scout, around 24 of which were unsuccessfully employed on anti-Zeppelin work in 1915.

Spirit
Former name of the Sikorsky S-76 twelve-seat civil helicopter, first flown in March 1977. The name was dropped in late 1980 to avoid offending certain religious groups in various (but unspecified) Latin American countries, and since then the only major variant of the series with a formal name has been the H-76 Eagle military export model first flown in 1985. Two S-76s have been modified to support the research effort for the Boeing/Sikorsky RAH-66 Comanche; the Fantail features a new enclosed tail rotor (broadly similar in appearance to Aerospatiale's Fenestron tail rotors) and the Shadow (Sikorsky Helicopter Advanced Demonstrator of Operator Workload) has a new single-seat cockpit grafted onto the nose of the standard S-76.

Spirit 750
Former name of Advanced Aircraft Corporation's Turbine P-210 conversion of the Cessna P-210 Pressurised Centurion lightplane, based around installation of a new 750shp P&WC PT6A-135 turboprop flat-rated at 450shp, and first flown in March 1984. A further turboprop conversion of the high-wing Centurion, with a 450shp Allison 250-B17F/2, is the Silver Eagle by O&N Aircraft of Pennsylvania, certificated in the US in late 1992.

Spoonbill
The Loening OL (US Navy) and OA-

SPRUCE GOOSE

2 (Air Corps) amphibious biplanes of the early twenties were both known as *Spoonbill*s or *Flying Shoehorn*s because of the seemingly upward-curved float projecting forward from under the fuselage.

Spraying Mantis
With three main rotors and a mass of struts and booms, the whole collection driven by a buried Rolls-Royce Merlin, the Cierva W11 Air Horse helicopter was one of the more unusual sights and sounds of the 1949 Farnborough flying display, and earned its best known nickname from its envisaged crop-spraying role. Despite its odd appearance, the *Clothes Horse*, as it was alternatively known, showed some promise, but only two were ever built.

Spruce Goose
Howard Hughes was said to have loathed the famous *Spruce Goose* nickname applied to his massive HK 1 or H-4 Hercules flying boat, and in fact the title was something of a misnomer as the sole example built (N37602) was constructed primarily from laminated birch. With eight engines and a span of 320ft (still the largest ever flown) the aircraft was originally intended to be the forerunner of a fleet of flying Liberty Ships to overcome the U-boat threat, but by the summer of 1947 it had still not flown and Hughes was given a public dressing-down by the Senate War Investigating Committee. A few weeks later, on the afternoon of Sunday, 2 November, before a crowd of 50,000 at Long Beach Harbor, Hughes himself was at the controls on that famous one-mile hop, after which it was mothballed in perfect condition for over 30 years, never to fly again. Following Hughes' death the aircraft

123

became a tourist attraction, alongside the Queen Mary liner. Debate continues over whether it was actually intended to fly that day, and in typically enigmatic fashion Hughes took the secret to his grave.

Staggerwing
Widely used but nonetheless unofficial name for the classic Beech Model 17 cabin biplane, also known as the *Flying Backstagger* because of its negative-staggered wings, i.e. the upper-wing further aft than the lower. The military communications variant was the C-43 Traveller.

Stanley Steamer
Such was the height and bulk of the Blackburn & General Universal Freighter Mk 1, that on its maiden take-off at Brough on 20 June 1950, chief test pilot Harold 'Timber' Wood turned to his co-pilot and casually enquired: 'My side's airborne; is yours?' An 80 per cent re-design (mainly to the fuselage) produced the Universal Freighter Mk 2, but whilst plans for civil car ferries with that name all foundered, the *Flying Cathedral* went on to play an invaluable role with the RAF as the Beverley, which name it received in December 1952. A total of 47 *Bevs* served faithfully until replaced by the Hercules from 1967, by which time their lumbering and archaic appearance had earned the fond nickname of *Stanley Steamer*.

Star Strutter
Well-known nickname used by both sides in WW1 for Austro-Hungary's Hansa-Brandenburg DI scout, due to the unusual configuration of interplane struts; *Spider* arose the same way but was much less common, as was *Coffin*, which referred to both

the deep, boxy fuselage and the DI's marked instability.

Sticking Plaster Bomber (i)
Nickname of the Luftwaffe's Focke-Wulf Fw 58 Weihe (Kite) twin-engined, trainer/comms machine, when used for casualty evacuation duties; alternatively *Leukoplastbomber* after the German brand of plasters.

Sticking Plaster Bomber (ii)
In contrast to the Weihe (see previous entry), the aptly-named Messerschmitt Me 323 Gigant (Giant) earned this title with the Afrika Korps because of its chronic vulnerability to fighter attack; the point was driven home on 22 April 1943, when no less than fourteen Gigants were shot down over the Gulf of Tunis. With a span of 180ft the Me 323 was easily the biggest transport of WW2; it was a six-engined derivative of the Me 321 glider designed for the planned Operation Sealion, the invasion of Britain.

Stingray
Resembling a less pugnacious-looking *Shagbat* with the engine mounted tractor-fashion on the upper wing, the Supermarine Sea Otter amphibian of 1938 was originally to have been known as the **Stingray**.

Flying Stone
Generic term for NASA's series of air-dropped, lifting-body research aircraft in the sixties and early seventies, notably the Martin Marietta X-24 (cf *Flying Potato*), and Northrop HL-10 and M2-F2 (cf *Cadillac*). Lifting-bodies were essentially wingless craft which derived their lift from the shape of their fuselages, and were seen as the possible forerunners of spacecraft

re-entry vehicles capable of landing on conventional runways, in a similar manner to today's Shuttle Orbiter.

Stoneboat
Canadian Bush Pilots' name for the sturdy de Havilland Canada DHC-3 Otter utility transport, originally known by its manufacturers as the King Beaver in view of its likeness to the smaller DHC-2 Beaver. The best known of a handful of turboprop DHC-3 conversions is the Vardax/Vazar Dash 3 (a title presumably inspired by the Dash 7 airliner) certificated in the US in 1989 and powered by a P&WC PT6A-135 instead of the original P&W R-1340 radial.

Stoof
For some time after its maiden flight in December 1952 the Grumman Tracker Anti-Submarine Warfare (ASW) aircraft continued to be known officially as the Sentinel, but the original pre-1962 designation of S2F soon gave rise to the popular nickname of *Stoof* in the US Navy. Similarly, the WF-2 (E-1) Tracer Airborne Early Warning (AEW) model with its huge dorsal-mounted radome became the *Willy Fudd* or *Stoof with a Roof*, but unfortunately the TF-1 (C-1) Trader Carrier Onboard Delivery (COD) derivative appears to have escaped the treatment.

The Firecat is a dedicated firebombing conversion of surplus Tracker airframes by British Columbia-based Conair, featuring single-point refuelling and a capacity for up to 3,295 litres of water or retardant; first deliveries were to the French Securité Civile in 1982, and the Turbo Firecat with new P&W PT6A-67AF turboprops flew for the first time in August 1988.

Flying Storevan
One of the unsung feats of between-the-wars aviation was when, in the twelve months from April 1931, four Junkers G31 trimotors flew the components of three complete dredgers (as part of a staggering 2,500 ton total uplift) into the New Guinea gold fields.

Flying Stovepipe
Contemporary US term for early jet aircraft. In the RAF there are those who have adopted the expression Blow Job to distinguish jet engines and jet planes from those propelled by airscrews, the classic example being the boast of Shackleton crews to their Nimrod counterparts that 'eight Screws will always beat four Blow Jobs . . .'

Stranny
Diminutive form of Supermarine Stranraer, the twin-engined RAF biplane flying boat that first flew in 1934. RCAF nicknames included *Strainer*, and *Whistling Bird Cage* from the humming of the bracing wires during flight.

Strat
Although it was built in modest numbers (56) compared to today's 'aluminium tube' airliners, the Boeing 377 Stratocruiser literally bulged with character; a two deck double-bubble fuselage, unprecedented passenger luxury, downstairs lounge bar (hence *Statuscruiser* and *Stratoboozer*), bunk beds, and a curious habit of nosewheel-first landings.

Derived from the B-29 Superfortress bomber, the *Strat* and its military counterparts the KC-97 tanker and C-97 Stratofreighter were also the basis of the bizarre Guppy freighter conversions with various wing-fuselage stretches and massively expanded upper fuselage lobes, originally conceived by Jack Conroy and Aero Spacelines for transporting NASA launch vehicle stages. The original conversion, carried out by On Mark Engineering in 1962, had a 19ft 9in diameter fuselage with a capacity of 29,170 cubic ft, and was soon nicknamed *Pregnant Guppy* after the aquarium fish; the name was later formalised, and Guppy became a generic title for all such conversions. The B-377PG was joined in 1965 by the largest of the series, the nose-loading B-377SG Super Guppy, with new P&W T-34 turboprops and a 25ft diameter fuselage of 49,790 cubic ft (for comparison roughly ten times the volume of a C-130 Hercules). The 1967 B-377MG, the so-called Mini Guppy, was closer in size to the original *Pregnant Guppy* but with a swing-tail instead of the earlier detachable unit, and the Guppy 101 was a further development along these lines but with nose-loading and new Allison 501 turboprops; sadly the 101 had a life of just two months, crashing on 12 May 1970. After these one-offs came the quartet of Guppy 201s (two of which were converted in France by UTA) powered by Allison 501 turbo-props and with a 39,000 cubic ft volume, well-known for their work with Airbus Industrie, who in fact purchased the rights in 1980. (For other variations on the Guppy theme cf *Beast, Whispering Giant* and *Quiet Giant*.)

Streamer
It is perhaps ironic that British Aerospace's successful Jetstream twin turboprop was originally conceived by Handley Page in an attempt to stay viably independent of the large industrial groupings of the day, i.e. BAC and Hawker Siddeley. Instead,

rising costs forced a reconstituted Handley Page Aircraft Ltd into liquidation in February 1970, and the design passed via Jetstream Aircraft Ltd (based at Sywell) and Scottish Aviation, to BAe, the direct successor to BAC and Hawker Siddeley.

The basic commercial version of the original HP 137 Jetstream was powered by Turbomeca Astazou turpoprops (as are the RAF's Jetstream T1s, built by Scottish Aviation), but before it went under the company also flew a prototype of the intended USAF C-10 (Jetstream 3M) model with Garrett AiResearch TPE731 engines. A number of HP aircraft operated in America by Apollo Airways also had their Astazous replaced with Garrett engines by Volpar as the Century Three Jetstream, and Riley Aeronautics converted at least two others to P&WC PT6A-34 power. BAe began develop-ment of its Jetstream 31 in December 1978, designating it thus because of its TPE731s, and the uprated Super 31 replaced it from September 1988. Apart from four Jetstream T3s for the Royal Navy and two for Saudi Arabia, most BAe aircraft were sold as small airliners, and to relaunch it on the corporate market in 1992 the company announced the nine-seat 'Formula 1' and twelve-seat 'Formula 2' Grand Prix Jetstreams, dubbed Corporate and Business Shuttle respectively for the USA; these were intended as marketing names to supplement rather than replace the established Jetstream title, but little has been heard of them since. The 29-seat Jetstream 41 is a stretch of the 31 (and thus a descendant of the original Handley Page design) but the Jetstream 61 is actually the former ATP (Advanced Turboprop) which evolved from the Avro/HS/BAe 748 (cf *Budgie*).

Stringbag

As with several other famous nicknames, there are conflicting theories regarding the origin of *Stringbag*, so widely used for the illustrious Fairey Swordfish naval torpedo biplane. Some say it was because of its ability to carry so much, others that it was due to its antiquated appearance by WW2, and Eric Partridge (quoting Hunt & Pringle) offered the original form of 'Stringbag the Sailor' – presumably a corruption of Sinbad. More certain, but equally unofficial, is *Blackfish*, relating to aircraft built by Blackburn at Sherburn in Elmet; the term is believed to have originated at Fairey's but many Service pilots also adopted it and claimed to be able to tell Fairey and Blackburn machines apart from the way they handled. Blackburn built 1,699 Swordfish (out of a total of 2,391) including both examples with the Royal Navy Historic Flight.

Stud

The most powerful helicopter outside of the Soviet Union/CIS is the suitably brutish-looking US Navy CH-53 Sea Stallion (Sikorsky S-65) or Super Stallion (S-80) in its three-engined CH-53E form. The US Marines' nickname of *Stud* has an obvious symbolic connection with the Stallion name, and the unfortunate title of *Super Shitter* denotes its habit of leaking hydraulic fluid, but there is also the more genteel *Buff* and the Vietnam-vintage *Super Bird*. The MH-53E model, recognisable by its enlarged sponsons and used for minesweeping with a towed 'bird', is formally the **Sea Dragon**, although in practice the name is often disregarded in favour of the usual Stallion name.

Early USAF HH-53s were quickly nicknamed *Super Jolly Green Giant*

after the smaller HH-3E (cf **Sea Dragon**) and over the years the title has become more or less formalised in the more manageable form of Super Jolly. Combat Rescue HH-53Cs were at one stage nicknamed *Black Knights*, but as the fleet has taken on more of a 'Special Ops' role they have become more closely associated with the Pave Low title of their advanced low-level navigation packages; the latest version to enter service, in 1987, is the MH-53J (Pave Low III enhanced). Amongst foreign operators of the S-65/CH-53 is Israel, where it is also known as the Yassour (Petrel).

Stuka

The name that was on everyone's lips during the early part of WW2 was actually an abbreviation of Sturzkampsflugzeug and thus, strictly speaking, valid for any dive-bomber. But of course *Stuka* will forever be synonymous with the Junkers Ju 87 and the never-to-be-forgotten sight of ugly, gull-winged machines diving vertically on their targets, accompanied by that terrifying scream from their 'Jericho Trumpet' sirens. Like so many of its Luftwaffe contemporaries, the Ju 87 was first put through its paces in Spain, where the Condor Legion took up Gunter Schwarzkopf's name for it of *Jolanthe the Pig* after a popular Berlin comedy play, 'Krach um Jolanthe ('Trouble for Jolanthe') starring a real pig; in contrast, Italy's Regia Aeronautica gave the Ju 87 the surprisingly flattering nickname of *Pichiatello* (*Little Woodpecker*). The anti-armour Ju 87G-1 variant, mounting a pair of underwing 37mm Flak 36 guns, was given the predictable but utterly Teutonic-sounding nickname of *Panzerknacker* or *Tank Cracker*.

Flying Submarine
Intended for the 1929 Schneider race,
Italy's novel Piaggio P7 monoplane
dispensed with the usual draggy floats,
and instead was fitted with a small pair
of hydrovanes to raise the aircraft up
from its semi-submerged state once
the ventral water-screw had propelled
it to sufficient speed; at that point
power would be switched to the
airscrew for a normal take-off. A top
speed of 373 mph was envisaged, but
the project was sadly abandoned with
insurmountable clutch problems.

Flying Suitcase
The nicknames inspired by the
distinctive outline of the Hampden
bomber (and its lesser sibling, the
Napier Dagger-powered Hereford)
are richly evocative of phoney war
atmosphere, of the *Spotter* magazine,
bike rides to the local airfield, and
those wonderful Wren
'Oddentification' caricatures and
rhymes: 'Is this what makes Herr
Hitler rage – a Hampden built by
Handley Page?' As seen from various
angles the Hampden was
affectionately described as the *Flying
Suitcase, Flying Tadpole, Flying
Panhandle, Frying Pan*, and even
(from the plan view) as the *Champagne
Glass. Hambone* was simply a play on
Hampden Bomber, and among the
formal names considered were
Huntley, Havoc, Hotspur, Harrier,
Halifax and Hastings.

Super Courier
Original title of the Airspeed AS45
advanced trainer, flown in prototype
form in 1941 as insurance against a
possible shortage of Miles Masters and
North American Harvards; despite
the name the AS45 had virtually
nothing in common with the pre-war
AS5 Courier civil monoplane, and the

title was later changed to Cambridge
to complement the same company's
famous Oxford trainer.

Super Hercules
Very early name for the Lockheed C-
141 StarLifter strategic jet freighter,
presumably because it has the same
basic 10ft 4in fuselage cross-section as
the C-130. Use of that cross-section on
the C-141 meant that on many
missions the hold would be full before
the maximum payload weight had
been loaded, so to help alleviate the
problem the USAF fleet of C-141As
was stretched by over 23ft from 1977
to become C-141Bs; in camouflaged
form these have been nicknamed
Conger Eels or *Starlizards*. The early
nickname of the *T-tailed Mountain
Magnet* for the C-141 seems to have
been a little harsh.

Super Hog
Despite the designation, virtually all
the Republic F-84F Thunderstreak
and RF-84F Thunderflash had in
common with their 'plank-winged'
Thunderjet forebears (cf *Hog*) was an
early reputation for sluggish handling,
which was largely overcome by the
introduction of a slab tailplane after
the first 275 aircraft. Another type
sharing the same broad designation
(though again with limited
commonality) was the XF-84H,
dubbed *Thunderscreech*,
Thunderprop and the *Mighty Ear
Banger* amongst other more profane
titles. Powered by an Allison XT-40
turboprop, the intense noise of the
supersonic propeller was said to cause
'acute nausea' to anyone within
several hundred feet of a ground-
running XF-84H (cf *Thud*).

Superhook
Vietnam-vintage nickname for the

Sikorsky S-64 Skycrane helicopter, known more formally to the US Army as the CH-54 Tarhe. The turbine-engined S-64 has its origins in the one-off piston S-60 Skycrane (N807) of 1959, and like its predecessor features a skeletal fuselage to enable it to operate either as a flying crane helicopter or to carry detachable, self-contained cargo pods. The Army used the CH-54 in Vietnam primarily for the recovery of downed aircraft, and a stripped-down *Super Lifter* crane version was also devised locally by removing the armour plating and the four cargo pod attachment points.

Super Super Sabre
Obvious nickname for the North American F-107, a Mach 2 fighter-bomber development of the F-100 Super Sabre (and initially designated the F-100B) distinguished by its unusual dorsal air intake. Although the three YF-107As performed well, the project was cancelled in March 1957 in favour of Republic's F-105 *Thud*.

Super Trader
One of the few beneficiaries of the miserable Avro Tudor piston airliner affair was Freddie Laker's Aviation Traders Ltd, who in September 1953 bought fourteen of the unwanted short-body Mk 1s and 4s as a job-lot, and after modification successfully put several back into service with associated company Air Charter, five as Super Trader long-range freighters. The basic Trader name had already been used by Avro for a planned Type 711A freighter conversion of the long-bodied Tudor Mk 2. The second prototype Tudor Mk 1 was fitted with four Rolls-Royce Nene jets in underwing pairs to become the Mk 8 in 1948, and it was followed by six

similarly powered Avro 706 Ashton pure research aircraft, built from components of the long-body Tudor Mk 2 but with a shortened fuselage and tricycle undercarriage. The Ashtons gave useful if unspectacular service for a number of years, and in 1960 WB493 also appeared in the film *Cone of Silence* as the Atlas Phoenix airliner. At the time of filming, WB493 was fitted with a Bristol Siddeley Olympus engine on the port wing and an Orpheus on the starboard, as well as its usual paired Nenes, all of which would have made the Phoenix one of the world's more noticeable jet airliners . . .

Sutlej
Indian Air Force name (after the Punjab river) for the Soviet Antonov An-32 (NATO – Cline) twin-turboprop tactical transport. The An-32P Firekiller variant has tanks on either side of the fuselage capable of carrying up to 17,635lb of water or fire retardant, and also carries flare packs to produce rainfall over the fire in a similar manner to the An-30M **Sky Cleaner**. The Firekiller made its first Western appearance at the 1993 Paris Show, but few are thought to have been built.

Swallow (i)
Initial name for Britain's first all-metal aeroplane, the one-off Short Silver Streak biplane of 1920. Although its structure was more akin to existing methods than the definitive stressed-skin formula that was to emerge later, the Silver Streak represented a bold attempt to move away from wood and fabric materials, and deserved better than the sceptical and even hostile approach it received from the Air Ministry.

Swallow (ii)
Anglicised version of the low-powered and economical Klemm L25 German monoplane of 1927, over 130 of which were built by the specially-formed British Klemm Co. at Hanworth, Middlesex, and its successor the British Aircraft Manufacturing Co. (BA).

Swallow (iii)
Appropriate but unofficial name for the de Havilland DH 108 tailless research aircraft, marrying what was essentially a Vampire fuselage to new, long, swept wings. In contrast to the pattern of many unofficial titles, the name was used informally by the under secretary of the Ministry of Supply, and it was de Havilland management who declined to endorse it, although shop-floor personnel frequently used the name. Three DH 108s were built but all were lost in crashes, including that on 27 September 1946 which cost the life of Geoffrey de Havilland Jr. On the credit side the trio provided much useful data on the behaviour of swept wings at both high and low speeds, and on 9 September 1948 John Derry made Britain's first supersonic flight in the third aircraft, albeit in a dive.

Swift
Upgraded version of the Romanian ICA IAR-99 Soim (Hawk) tandem-seat jet trainer/light-attack aircraft, by Israel Aircraft Industries. Designated IAR-109, the Swift features new computers, a new Head-Up Display (HUD) and a new steerable nosewheel. Flight tests began in Israel in November 1993.

Syuan Fen
Syuan Fen (Whirlwind) was the name applied by China to the civil version of its H-5 helicopter, a licence-built version of the Soviet Mil Mi-4 (NATO – Hound).

Tabloid
As with several other Sopwith types, the *Tabloid* name applied to their compact and sprightly two-seater of 1913 was never formally adopted by the company, and yet became much more commonly used than many carefully chosen official titles. The name was taken from that of a popular small first-aid kit because, said the *Aero*, 'it is a concentrated dose of medicine for certain gentlemen at the Royal Aircraft Factory'. To its makers and (formally at least) the RFC, it was usually just the scout (arguably the first use of that term), but even so the manufacturers of the first-aid kit threatened legal action over use of the *Tabloid* name, the only result of which was to prove, yet again, that the most effective way of establishing any nickname is to try to suppress it. Production aircraft were single-seaters, and as well as winning the Schneider Trophy at Monaco on 20 April 1914, the *Tabloid* was also used by the RNAS in the historic raid on the Zeppelin sheds at Dusseldorf on 8 October 1914. Mainstream developments during WW1 were the Schneider and Baby seaplanes for the RNAS, the latter distinguished by a more conventional horseshoe style cowling in place of the earlier 'bull's-nose' type.

Tamiz
E26 **Tamiz** (Grader) is the Spanish Air Force designation for the Chilean ENAER T-35 Pillan basic trainer, developed by Piper and built largely around **Cherokee** and Saratoga components. Pillan is pronounced 'Pee-Yan', and means 'devil' or 'demon' in the dialect of the

Araucanos Indians of southern Chile. The Allison-powered Aucan (Blithe Spirit) converted by Soloy in the US was later renamed Pillan Turbo.

Tarzan
One of the last throws of the long-outdated *Shturmovik* concept, the single-turboprop Tupolev Tu-91 (NATO – Boot) first revealed to the West in 1956, was supposedly known to its test pilots as both the *Tarzan* and *Bichok* (*Steer*).

Taube
When the licensing agreement broke down between the Austrian designer Igo Etrich and the German Rumpler concern, numerous German companies took up the basic design of the 1910 **Taube** (Dove) monoplane, so named because of its characteristic curved wings and dovetail-shaped empennage. Whilst staying with the basic formula, various companies made their own modifications and thus the title came to be regarded as a generic term rather than a proper type name for any single brand (cf *Invisible Aeroplane*).

T-Bird
Whilst Americans, in particular, sometimes use *T-Bird* for any kind of trainer, to those of a certain age, or those with a sense of history, the name is synonymous with the Lockheed T-33. Finally retired from the USAF in 1988, the ubiquitous T-33 was a pure trainer derivative of the F-80 Shooting Star (America's first really practical jet fighter), and other mainstream variations on the theme were the Navy's T2V (later T-1) SeaStar trainer with stepped cockpit and leading-edge slats, and the Air Force's F-94 Starfire series of two-seat, all-weather fighters. Among numerous foreign

operators was the Japanese Air Self-Defence Force (JASDF) which in 1964 belatedly assigned the name Wakataka (Young Hawk) to its mainly Kawasaki-built fleet. Canada ordered twenty T-33As with the normal Allison J33 engine from Lockheed as Silver Star 1s in 1951, plus a single Silver Star 2 adapted to take a Rolls-Royce Nene; this latter machine served as the pattern for no less than 656 T-33AN (later CT-133) Silver Star 3s built under licence by Canadair.
Easily the most unusual T-33 variant was the Skyfox, a prototype of which was converted for Skyfox Corp. by Flight Concepts Inc. and first flown in August 1983. With twin nacelle-mounted Garrett TFE-731 turbofans, new tail and rear fuselage, leading-edge root-extensions (LERXs), downturned winglets and a fatter nose, the venerable old **T-Bird** was almost unrecognisable, despite claims of 70 per cent structural commonality. OGMA of Portugal was slated to undertake quantity conversion and in 1986 Boeing Military Airplane Co. acquired the rights, but the project failed to find a market.

T-Bone
Beech evolved their Model 50 Twin Bonanza six-seat light twin by marrying a widened Model 35 Bonanza fuselage to the tail and outer wings of the T-34 Mentor military trainer. The first flew in November 1949, and as well as widespread civil sales it was adopted by the US Army as the L-23 (later U-8D and E) Seminole liaison aircraft, and later also as the RL-23D with Side-Looking Airborne Radar (SLAR) for radar surveillance. Excalibur was the name used by Texas-based Swearingen Aircraft for a conversion applicable to

the D50 and subsequent models of Twin Bonanza, with new 380hp Lycomings, fibreglass nose, new undercarriage doors, and built-in airstairs; 26 conversions took place, plus ten Excalibur 800s with 400hp engines, before the programme was acquired by sub-contractors Excalibur Aviation (cf **Queenaire**).

T-Cat
Shorthand term for the US Navy/ Marine Corps F7F Tigercat, the twin-engined Grumman fighter which arrived just too late for operational use in WW2.

Tebuan
Royal Malaysian Air Force name (after an indigenous species of wasp) for the twenty Canadair CL-41G Tutor armed jet trainers delivered in 1967.

Teen Series
In contrast to the earlier USAF *Century Series*, the *Teen Series* generation included both Air Force and Navy fighters, namely the Grumman F-14 Tomcat (cf *Turkey*), McDonnell Douglas F-15 Eagle (cf *Twin-tailed Tennis Court*), General Dynamics/Lockheed F-16 Fighting Falcon (cf *Viper*), and McDonnell Douglas F/A-18 Hornet (cf **Cobra**). Nothing is known of any F-13, the Northrop YF-17 (cf **Cobra**) lost the 1975 ACF fly-off against the F-16, and although the F-19 designation was widely attributed to the Lockheed 'Stealth Fighter', that of course finally emerged in November 1988 as the F-117 (cf *Wobblin' Goblin*).

Tempo
As well being one of the best-looking aircraft of WW2, the Douglas A-26 (B-26 from 1948) Invader had a

performance that made surplus aircraft ideal for civil executive conversions. Perhaps the best known was the ten- to thirteen-seat **Tempo** by L.B. Smith of Miami, with a new but superficially similar fuselage over 9ft longer than the original, tip tanks, and new wing attachments to remove the spar from the cabin. The Tempo I was the unpressurised model, but the first to fly in late 1959 was the pressurised and more powerful Tempo II; yet a further stretch produced the Biscayne 26, named after a district of Miami. The Lockheed Air Services Super 26 also featured an enlarged fuselage, plus airstairs and Constellation cabin windows. Rhodes Berry of Los Angeles circumvented the problem of the obtrusive spar by deepening the lower fuselage underneath it, giving their 1960 vintage Silver Sixty a 6ft 6in headroom and space for up to fourteen seats; a cargo version with rear-loading ramp was also mooted, but appears not to have been built. The Monarch 26 by the Aircraft Division of Rock Island Oil was much simpler, with new interiors and avionics in a basically standard B-26 airframe.

California-based On Mark Engineering had the most successful of all B-26 conversions with their Marketeer and pressurised Marksman, both featuring a re-designed cockpit, DC-7 windscreen and cabin windows, and a circumferential spar in place of the standard straight-through unit. On Mark's experience with the type also led to the B-26K Counter Invader COIN (Counter Insurgency) conversion for USAF Special Operations in Vietnam, with quickly interchangeable eight-gun or glazed noses, eight underwing hardpoints, and deleted ventral and dorsal turrets.

The first of 40 'production' B-26Ks
(also known as the A-26, for political
purposes) flew in May 1964, and until
fatigue finally overtook it the type
enjoyed considerable success over the
Ho Chi Minh trail at night, operating
out of Nakhon Phanom ('Naked
Fanny') Air Base in Thailand.

Ten
Title applied by Avro to the fourteen
Fokker F VIIB/3m tri-motor
monoplanes it built under licence from
1929; the name referred to the fact
that it carried eight passengers and
two crew, and should not be mistaken
for its Avro type number (618).

Tercel
Tercel (a male falcon) was the winner
in a Hawker Siddeley competition to
find a title for the new HS 1182
advanced trainer, but was rejected by
the RAF in August 1973 in favour of
the runner-up name, Hawk; one of the
reasons was that a Hawk is the symbol
of the RAF Staff College. The US
Navy's carrier-capable derivative,
developed jointly by BAe and
McDonnell Douglas, is the T-45
Goshawk.

Terminator
Consolidated always wanted to call
their B-32 bomber (built as a back-up
to the Boeing B-29 Superfortress
programme) the Terminator, but in
late 1944 the Technical sub-
Committee on Naming of Aircraft
decided on Dominator instead.
However, in August 1945 the State
Department objected (presumably it
was considered inappropriate for the
post-war era) and Dominator was
dropped; that left the B-32 officially
nameless, although Dominator is still
freely used, and there are also those
who claim that in lieu of a formally

ratified replacement name,
Terminator still stands.

Three Finger Bomber
The supremely adaptable Junkers Ju
88 was known in German recognition
manuals as the *Three Finger Bomber*,
which neatly describes the close
grouping of the nose and engines, and
the Ju 88P-1 conversion with a 75mm
PaK 40 anti-tank cannon in a huge
belly fairing was dubbed *Fat Bertha*.
In late 1944 the improved Ju 188 series
was given the belated and now almost
forgotten title of Racher (Avenger),
and around the same time the Ju 388J
night-fighter (but no other 388
models) was named Störtebeker after
a mythological German pirate figure
(cf **Misteltoe**).

Three-Holer
Though it was by no means the world's
only tri-jet (or even the first), the fast-
selling Boeing 727 airliner really made
the *Three-Holer* title its own. Eastern
Air Lines also knew the 727 somewhat
optimistically as the Whisperjet; to
Lufthansa it was the Europa Jet, and
Trans Australia promoted it as the
T-Jet.

Thud
In the early days its detractors (and
there were many) dubbed the F-105
Thunderchief strike-fighter the
*Heavyweight Glider, Iron Butterfly,
Lead Sled* and even *Squash Bomber*,
saying that the only way it would ever
destroy anything was by taxiing over
it. As Republic's own successor to the
F-84 *Hog* series it was also known as
the *Ultra Hog* or *Hyper Hog*, but
Super Hog really belongs to the swept
F-84s. The classic nickname of *Thud*
originated as *Thunderthud*, yet
another jibe at the size and weight of
the beast, and contrary to some

accounts appeared long before the
Vietnam War. There, it was
sometimes claimed to represent the
sound of all the hits the tough 105s
took, and the standard bar-room joke
among F-4 Phantom pilots was:
'What's the sound of an F-105
augering-in (i.e. crashing)?' Answer:
'Wheee . . . Thud!' In time, though, the
name came to symbolise the respect
held for the aircraft which by the
beginning of 1968 had mounted no less
than 75 per cent of all USAF missions
over North Vietnam, and was on its
way to downing 27½ MiGs. The F-105
was so heavily committed to the strike
effort up north that a karst line there
went down in USAF folklore as Thud
Ridge from the seemingly endless
procession of F-105s flying along it on
strikes in the heavily-defended Hanoi
area; with typical black humour this
was known as 'going downtown', after
the 1964 Petula Clark song which
promised 'downtown . . . everything's
waiting for you'. As a consequence,
the air force lost nearly half the F-105
inventory in south-east Asia; almost
400 out of a total of 833 built.

Thunderstick (or T-Stick) was
Republic's name for the AN/ASG-19
Fire Control System of the F-105D,
but the title is perhaps best
remembered for the AN/ARN-92
Thunderstick II navigation kit later
fitted in a prominent dorsal spine to 30
F-105Ds (cf **Wild Weasel**).

Thunderbolt Amphibian
Representing a rather blatant attempt
to capitalise on the fame of their P-47
Jug, **Thunderbolt Amphibian** was the
misleading title applied by Republic
Aviation to their 1944 RC-1 private-
owner light amphibian, based on the
general layout of the pre-war Spencer
Air Car. The production model with a
new wing and simplified structure was

the RC-3 Seabee, named after the
famous wartime Marine Corps
Construction Battalions (CBs or
'Seabees'). Republic's venture into
civil aviation proved a brief and
unhappy affair, despite building over
1,000 Seabees up to 1947. Among
projected but unbuilt variants were
the Landbee without the usual
amphibian hull, the two-seat Beebee
trainer using numerous RC-3
structures, and the Twinbee with
Franklin motors paired to drive the
existing single pusher prop. The latter
should not be confused with the mid-
sixties-vintage Twin Bee conversion
by United Consultants, with twin
Lycomings or Continentals mounted
on the wings in more conventional
tractor fashion, plus a fifth seat in
place of the old engine compartment;
at least 24 conversions are known.

Thunder Piglet
The enduring power of aviation
nicknames and colloquialisms was
perhaps never better illustrated than
by the case of the Fairchild T-46 twin-
jet trainer, christened *Thunder Piglet*
by the traditionalists who remembered
the illustrious line of Republic
Thundercraft and *Hogs* which had
preceded it. Sadly, all the traditions of
Seversky, Republic and Fairchild were
to end not with a bang but with a
whimper; the badly mismanaged T-46
programme was cancelled in March
1987, and the Farmingdale 'Thunder
Factory' was finally silenced soon
after.

Thunderwarrior
Name unofficially earmarked for the
advanced Republic XF-103
interceptor, cancelled in August 1957
before the first aircraft was
completed. Powered by a Wright dual-
cycle turbo-ramjet engine, and with a

projected speed of Mach 3.7 at 100,000ft, the F-103 would presumably have finally shaken off the *Hog* label which had haunted its Republic predecessors.

Tiffie

Popular nickname for the RAF's pugnacious Hawker Typhoon fighter-bomber of WW2, powered by the temperamental Napier Sabre of up to 2,260hp; *Bomphoon* was a much less common term, and probably dates from the switch from fighter to close-support operations around 1942. A parallel development with a Rolls-Royce Vulture engine, the Tornado, actually flew before the Typhoon in October 1939 and was regarded as the more promising of the pair, before engine problems led to its abandonment.

Tiger

Well-known name used by British helicopter company Bristow for the Aerospatiale/Eurocopter AS 332L Super Puma. Puma adaptations known as the XTP-1 Beta and the Gemisbok were used by South Africa's Atlas Aircraft to develop the avionics and weapons systems for the CSH-2 Rooivalk combat helicopter; first flown in 1990, the Rooivalk (Rock Kestrel) uses Puma dynamics in a completely new airframe, and an initial batch of sixteen was ordered for the SAAF in March 1994. Atlas has also converted a number of SAAF SA 330 Pumas into the more powerful Oryx by installing the engines and rotor system of the SA 332 Super Puma. Eurocopter's current military export model of the Puma is designated AS 532 Cougar, and has its commercial counterpart in the AS 332 Super Puma II.

Tiggie

Affectionate term for the de Havilland DH 82A Tiger Moth, the much-loved trainer biplane on which so many WW2 British and Commonwealth pilots first learned to fly. Perhaps the best known conversion applied to the Tiger was the widebody Jackaroo (a kind of Australian sheep-farm worker) with two pairs of side-by-side seats in an enclosed cabin. Nineteen were converted by Jackaroo Aircraft at Thruxton, Wilts. (hence the common name of Thruxton Jackaroo), including an unsuccessful single-seat agricultural version; a single Jackaroo was also converted by Rollasons at Croydon.

Timmy

Occasional RAF nickname for the Lockheed TriStar tanker/transport, around the time it entered service with the re-formed 216 Squadron in 1983/84. The two aircraft deployed to Saudi Arabia during the Gulf War with 'sand pink' camouflage were known collect-ively as Pink Pigs and individually as 'Pinky' (ZD949) and 'Perky' (ZD951). In its more familiar role as a widebody commercial airliner, the L-1011 TriStar was given the name Whisperliner by Eastern Air Lines, and when it first entered service with BA it was allegedly known as the *Big Dipper* because of the increased opportunity for transferring complimentary First Class spirits into miniature bottles ('dipping') which were then sold on a private-enterprise basis to Economy Class passengers. BiStar was the name attached to a proposed twin-engined version seating 200-250, studied around 1972.

Tin Donkey

Occasional nickname for early Junkers metal-skinned types, but most

135

closely associated with the pioneering J 1 monoplane, which became the world's first all-metal aircraft when it flew on 12 December 1915.

Tin Goose
Ford moved into aviation by acquiring the Stout Metal Airplane Company in 1925, and though they stayed in the industry just seven years (not counting WW2 'shadow' type production) their 4-AT and 5-AT Tri-Motor transports had virtually the same revolutionary effect on the American airline industry that their Model T car had on the auto industry. Although *Tin Goose* was by far the most popular name for the tough, dependable Fords, *Tin Lizzie* (from the Model T) was sometimes used, and of course *Flying Washboard* referred to the corrugated metal skin.
　　Several Fords continued to earn their keep into the sixties and seventies and there have even been attempts to introduce an updated version for production in developing countries. Renamed Bushmaster 2000 because Ford would not relinquish copyright in the original title, the new version featured a taller fin, new cowls to the previously exposed radial engines, and a deeper windscreen, whilst still retaining that purposeful Ford look. The first was completed by Aircraft Hydro-Forming in 1966 and a second by Bushmaster Aircraft Corp. in 1985, and despite the fact that the original design was then nearly 60 years old the Bushmasters proved the equal of modern turbine STOL types.

Tinker Toy
No other type can have been more deserving of the *Tinker Toy* label than the petite Temco TT-1 Pinto primary jet trainer; with just 920lb of thrust it stuck to a hot runway like chewing

gum, and the fourteen purchased for evaluation by the US Navy were placed in storage in 1958. American Jet Industries (AJI) flew a more powerful (2,850 lb st) T-610 Super Pinto trainer/counter-insurgency version in June 1968 and later sold the rights to the Philippine Air Force, whose Self-Reliance Wing planned to build it as the Cali (Hawk). The final incarnation was in the late eighties as the Avstar T-100 Super Mustang, with construction planned for Shenyang, China, and final fitting-out in the US but this project also foundered.

Tin Mossie
Commonly used nickname for the Vickers Type 432 high-altitude fighter prototype (DZ217) of 1942, although the resemblance to the illustrious DH Mosquito was quite superficial.

Tin Wimpy
Between them, *Tin Wimpy* (after the Wellington bomber) and *Pregnant Dakota* pretty well sum up the portly appearance of the post-war Vickers Viking piston airliner. Until it was replaced by the Viscount from 1953, the twin-engined Viking formed the backbone of BEA, and from 1952 the airline had most of its Viking 1B fleet modified to Admiral class standard with a 36-seat tourist-class layout in place of the earlier 27-seat first-class arrangement; as with the contemporary Pionair (Dakota conversion) the change was largely made possible by removing the wireless operator's station. Sixteen BEA aircraft were later sold to Eagle Aviation, who in July 1953 unveiled its 36-seat Mayfair class, also known as the Troopmaster when fitted with aft-facing seats for War Office trooping flights to the Mediterranean and Africa. The RAF counterpart was the

Valetta (the original spelling for the Maltese capital, with a single 'l') which was widely known in service as the *Pig*, particularly the T3 navigation trainer with a row of astrodome 'teats' along the cabin roof. A more extensive RAF development was the Varsity multi-role trainer with a new tricycle undercarriage, more neatly cowled engines, and an under-fuselage pannier housing a bomb-aiming station and a small bomb bay.

Tits Machine
American slang for a no-frills, hack-it fighter. Thus, the Northrop F-20 Tigershark would have been a *Tits Machine*, but the Grumman F-14 Tomcat is not; a BAC Lightning would be, but not a Tornado F3.

Tommy
The most successful indigenous American warplane of WW1 was probably the S-4 scout, built by the Thomas Morse Aircraft Corp. of Ithaca, NY and universally known as the *Tommy* or *Tommy Scout*; 100 S-4Bs and roughly 500 S-4Cs were completed, but none were used operationally in Europe. As well as appearing in numerous Hollywood epics in the twenties as pseudo-Camels, Nieuports, *Spad*s, etc., a number of post-war civil conversions were undertaken, usually involving replacement of the Gnôme Monosoupape or Le Rhône rotary engine. The best known was the two-seat Yackey Sport with a Curtiss OX-5 inline motor, by the Yackey Aircraft Co. of Illinois, but the Dycer Sport produced by Dycer Aircraft of LA was very similar.

Toofani
Name applied by the Indian Air Force to the Dassault MD 450 Ouragan

single-seat jet fighter, both French and Indian titles translating as 'Hurricane'. Dassault's first jet fighter, the Ouragan first flew in early 1949, and a total of 350 were built for the Armée de l'Air, 104 for India, and 24 for Israel, who also received another 51 surplus aircraft from France. A proposal for a two-seat night-fighter model with a new 'solid' radar nose, the MD 451 Aladin, progressed no further than a standard machine being fitted experimentally with new cheek-mounted intakes, and also limited to prototypes was a rough-field STOL variant with new twin-wheel main undercarriage units, named Barougan at the expense of its rival, the SNCASE (Sud-Est) SE-5000 Baroudeur.

Transporter
Little-used name for the Gates Learjet Model 25 corporate jet, circa 1966. Century III Learjets (introduced in Bicentennial Year, 1976, hence the name) were those incorporating a number of aerodynamic refinements, notably to the wing leading-edge; Longhorn was the former name of the 1977 Models 28 and 29 and the larger and more prolific 50 Series Learjets, which introduced Whitcomb winglets in place of the traditional tip-tanks.

Conceived in Switzerland by the American Bill Lear (inventor of the eight-track stereo) as the SAAC-23 but put into production at Wichita, the original Learjets used the same basic wing as the P-1604 *Jet Stork* fighter, and as a consequence had a reputation for being unduly 'hot' for a corporate aircraft (cf *Learstang*, under **Apache**).

Trenchard
Name rumoured to have been pencilled in for the controversial BAC TSR-2 strike aircraft prior to

cancellation in 1965. Hugh Trenchard (1873–1956) was Chief of the Air Staff between 1918 and 1929, and is of course remembered as 'The Father of the RAF'.

Tri-4
Early company title for the experimental Bell XP-77 lightweight fighter of 1944, reflecting a target of a speed of 400 mph and a gross weight of 4,000lb with a 400hp motor. The wooden XP-77 was conceived as a means of preserving strategic war materials, but delays and a disappointing performance meant it never entered production.

Tripehound
Few fighters have had such a brief but spectacular career as the Sopwith Triplane, which served the Royal Naval Air Service for just seven months in 1917 before giving way to the *Camel*. Famous as the mount of (amongst others) Raymond Collishaw's renowned B Flight of 'Naval Ten' squadron, the *Tripe* or *Tripehound* was famous for its phenomenal rate of climb, which someone once said gave it the appearance of 'an intoxicated staircase'.

Trixie
Pet name for the little General Aircraft GAL 55 training glider, derived from the Air Ministry TX3-43 specification from which it evolved; two only were built, flying for the first time in late 1943.

Trojan Horse
A one-off seaplane version of the Gotha GI twin-engined German bomber of WW1, the Gotha UWD earned its nickname by carrying six people on a trials flight from Warnemunde in 1916.

Troopship
Though little used in its own right, the Friendship name of the highly successful Fokker F27 turboprop airliner is another of those adaptable titles that could have been tailor-made for variations. As early as 1952 when the F27 was first announced (with optional piston engines, incidentally) the company used the parallel terms Friendship and Freightship for airliner and cargo versions, and **Troopship** later became well-known for the military F27M with cargo door and twin parachutists' doors. Long Friendship was a thankfully short-lived title for the stretched Series 500 circa 1962, and Combiplane was used by Fokker for the Series 300 and 400 mixed-traffic models. One development not proceeded with was the 1973 P305 Commuter, with Avco Lycoming ALF 502 turbofans (as used in the BAe 146) on pylons outboard of the existing nacelles, the rear portions of which would have been retained to house the undercarriage.

Other military models include the essentially unarmed Maritime of 1976 as sold to Peru (two), Spain (three), Netherlands Air Force (two), Angola (one), Nigeria (two), The Philippines (three) and Thailand (four with limited armed capability), and the armed Maritime Enforcer of 1984 with eight external hardpoints for torpedoes or sea-skimming missiles. Fokker also offered a Sentinel variant for overland surveillance, the Kingbird AEW aircraft with dorsal-mounted, phased-array radar, and the Black Crow for Elint/Sigint operations; all military versions are also available in Mk 2 form, using the airframe of the commercial Fokker 50, which differs from the F27 in its glass cockpit, composite structures, and P&WC PW125B turboprops in place

138

of the faithful Rolls-Royce Darts; a 64in stretch of this also produced the Fokker 60 Utility, four of which were ordered by the Royal Netherlands Air Force in the Spring of 1994.

Included in the overall production total of 786 (not counting Fokker 50s) are 205 F-27s and stretched FH-227s by American licensee Fairchild Hiller, and in fact the first of the family to go into service anywhere, with West Coast Airlines on 28 September 1958, was an American-built aircraft. Among various models tailored to the US market was the Cargonaut conversion, with a freight door fitted by American Jet Industries. The Firefighter is yet another conversion by Conair of British Colombia, carrying up to 6,364 litres of fire retardant dropped through computer-sequenced doors, and retaining full main-deck freight capability; the first was converted from a Dutch-built F27 Series 600, and received its type approval in June 1986.

Turbo Star 402 & 414
Turboprop conversions of the Cessna 402 Utililiner/Businessliner and 414 Chancellor twins with Allison 250 engines in place of the usual Continental piston units. Initiated by American Jet Industries (AJI) in 1969, the programme was sold in 1977 to Scenic Airlines of Las Vegas.

Turkey
When they announced Tomcat as the title for the new F-14 naval fighter in 1970 it seemed almost a parody of Grumman's long line of fighting feline names, but over the years 'Top Gun', Sukhoi and MiG killing, and the sheer firepower of the thing have all helped give the name a cachet that was probably unmatched among fighters until the arrival of the Su-27 Flanker;

all of which perhaps helps put into context the *Turkey* epithet used by drivers of F/A-18 Hornets and F-16N aggressors. An ambitious proposal to upgrade the F-14D to multi-role Quickstrike configuration with state-of-the-art avionics and engines was not taken up, but in 1992 fleet F-14s began to receive a strictly limited bombing capability as *Bombcats*. Tomcats with the TARPS reconnaissance pod are, naturally, *Peeping Toms*.

Tweenie
Generally known as the Churchill because it was sponsored by the then First Lord of the Admiralty, the one-off Sopwith two-seat biplane of 1914 was also referred to as the *Tweenie* (the contemporary term for a 'between floors' maid) as it came between the RFC's 80hp three-seater and the *Tabloid* in Sopwith's current family. *Sociable* is also used in connection with the aircraft, which as 'No. 149' served the RNAS as a trainer, but this was a more general term, popular at the time for aircraft with side-by-side seating.

Tweet
Serving as the USAF's primary/basic jet trainer since 1957, the Cessna T-37 has been fondly known to untold thousands of American aircrew as the *Tweety Bird* or *Tweet*; less common are *Squeak*, *Hummer*, *Baby Jet*, and the *Six Thousand Pound Dog Whistle*. **Dragonfly** is the name of the definitive armed A-37B with tip-tanks, eight underwing hardpoints, built-in minigun, and provision for inflight refuelling; production of 577 A-37Bs was preceded by a pair of YAT-37D armed prototypes in 1963 and 39 unnamed A-37As for use in Vietnam.

Twin Six
Original title of the Piper PA-34
Seneca light twin, reflecting its origins
in the single-engined PA-32 Cherokee
Six. From 1979 the Seneca II was also
built under licence in Poland as the
PZL-M20 Mewa (Gull) and from 1992
this was being marketed in the West
with the title M20 Gemini.

Twin Stinson
Piper inherited the basic design of its
trend-setting **Apache** twin in its
takeover of the old Stinson company
in 1948, and retained the prestigious
title for the twin-tailed, fabric-covered
prototype which flew in March 1952,
but by the time the definitive, all-
metal, single-tailed PA-23 flew in
December 1953 it had been renamed,
becoming the first of a long line of
Piper types to bear the names of
Indian tribes.
    In the early sixties the *Fat Albert*
was supplanted on the Piper lines by
the swept-fin Aztec development, and
in 1961 Vecto Aircraft of San
Antonio, Texas began offering their
successful Geronimo conversion,
named after the chief of the Apaches.
Improvements upon the baseline
Apache 160 included more powerful
180hp Lycomings, an extended
(almost Aztec-style) fibreglass nose,
new wingtips and a new one-piece
windscreen; at least 30 Geronimos had
been completed by 1965, and two
years later the programme was
resumed by Seguin Aviation. The
contemporary Dart II conversion by
Doyn Aircraft of Wichita also had
180hp Lycomings, but the 1969 Miller
Jet Profile featured 220hp Franklins as
well as a revised tail.

Twin-Tailed Tennis Court
Like its F-14 naval counterpart (cf
Turkey), the McDonnell Douglas F-15

Eagle air superiority fighter has two
major drawbacks – cost, and the fact
that in any air-to-air engagement its
size makes it stick out like the
proverbial sore thumb. The F-15E
Strike Eagle is also nicknamed *Mud
Hen*, and *Beagle* comes from 'Bomber
Eagle', although of course the F-15E
retains much of the air-to-air
capability of the C and D models. In
the Israeli Air Force the F-15 has the
alternative name of Baz (Falcon).

Twotter
Occasional nickname for the well-
known de Havilland Canada DHC-6
Twin Otter STOL transport. For
sightseeing flights through the Grand
Canyon, Scenic Airlines and Grand
Canyon Airlines both operate a
customised version known as the Vista
Liner, with enlarged windows and
quieter four-blade Hartzell props.

Ugly Luscombe
Jibe at the post-war era Cessna 120
and 140 lightplanes by aficionados of
the more classy Luscombe 8 Silvaire.

Ute
Just as the US Army's L-23F (U-8F)
had helped pave the way to the civil
Model 65 Queen Air, so its NU-8F of
1963, with new P&WC PT6A
turboprops, was instrumental in
developing the best-selling King Air
corporate aircraft. The first genuine
Model 90 King Air flew in January
1964, and was in turn adopted by the
Army as the U-21 **Ute**; the US Navy's
T-44 trainer is also based on the
original short-body Model 90 series,
and has the seldom used name of
Pegasus. Among a string of battlefield
Army Elint derivatives was the RU-
21H, converted under the Guardrail V
programme. Not to be confused with
Army programmes is the 1991

Raytheon (Beech's parent company) RC-350 Guardian, intended for export, and converted from the prototype King Air 350 (stretched, wingletted 300) with much of the ESM suite of the US Navy's Sikorsky SH-60B Seahawk helicopter. Taurus is the name of a conversion by Jetcrafters Inc. (formed by Ed Swearingen in Texas) which retrofits Garrett TPE-331 turboprops to the Model 90, and thus approximates to Beech's own King Air B100.

First flown in 1972, the Model 200 Super King Air is readily identifiable by its T-tail and increased span, and has its military counterparts in the multi-service C-12 series, known to the Army as the Huron; Elint versions are the RC-12D and H (Improved Guardrail V programme) and the RC-12K (Guardrail Common Sensor). The one-off Fan Jet 400 (alias King Air 400) of 1975 with overwing P&W JT15D turbofans was an attempt to introduce a corporate jet to the Beech line, prior to acquiring the **Beechjet** from Mitsubishi. The biggest stretch so far is the Model 1900C (military C-12J) Airliner seating nineteen and corporate Model 1900 Exec-Liner; first flight was in 1982, and the 1900D with stand-up headroom followed in 1990.

Vajra
Indian Air Force name for the Dassault Mirage 2000H fighter. The first **Vajra** (Thunderbolt) entered service with No. 7 Squadron IAF (the Battleaxes) in 1985.

V-Bomber
Contrary to popular belief, the V-Bomber title, so much a part of Cold War language, originated not with Sir Winston Churchill but with his Chief of the Air Staff, Sir John Slessor. The

Vickers Type 660 was named Valiant (cf *Black Bomber*) in June 1951, but at a meeting of the Air Council on 2 October 1952 Slessor effectively ended expectations of similarly alliterative names for the Avro 698 and Handley Page HP 80 (cf **Albion** and **Harpoon**) by expressing a preference for establishing a 'V' class of bomber (i.e. V-for-Victory) with Vulcan and Victor. Perhaps if Avro or HP had been first, we would have had A-Bombers or H-Bombers, which would have coincided with their nuclear role (?). A further misconception about the mark is that the Short SA4 Sperrin was in some way the 'missing' fourth **V-Bomber**, but in fact whilst the two prototypes (VX158 and 161) undoubtedly contributed to the programme the Sperrin (named after the mountain range in Northern Ireland) was developed to an earlier specification.

Vee Strutter
Widely used RFC/RNAS term for the much-respected Albatros DIII scout, first of the famous German fighter series to dispense with parallel interplane struts.

Veltro II
Former name (after the WW2 MC205 Veltro [Greyhound] fighter) for the Aermacchi MB-339K, a single-seat, ground-attack model of the MB-339C tandem-seat jet trainer; the sole example built to date made its first public appearance at the 1980 Farnborough show, wearing the appropriate Italian civil registration I-BITE. The MB-339 variant being promoted by Lockheed for the USAF/USN JPATS programme (Joint Primary Aircrew Training System) is labelled the **T-Bird II** after the famous Lockheed T-33.

Versailles
Name originally proposed for the late
WW1-vintage Vickers Vimy twin-
engined bomber; ironically, the
northern French city of Versailles was
later of course to give its name to the
peace treaty which put a formal end to
WW1.

Vespa Jet
Briefly used title of the Piaggio/
Douglas PD-808 corporate jet,
exploiting the worldwide fame of the
Italian company's Vespa motor
scooter. The first of four prototype
and development PD-808s flew in
1964 and was followed by a run of 25
for the Italian Air Force, but no
commercial sales were attained.

Vibrator
Widely used epithet for America's
wartime Vultee BT-13 Valiant trainer,
reflecting the more than adequate
warning it gave when about to stall.
The forerunner of the unloved
Valiant, the BC-51, was promoted by
Vultee as the *Eighty-Eight Day Plane*,
after its rapid transition from drawing-
board to first flight in 1939.

Vic
Airmen of the inter-war RAF
apparently declined to differentiate
between the long-serving Vickers
Victoria transport biplane and its
higher-powered and beefed-up
derivative the Valentia, referring to
both by the diminutive *Vic*. Of the 82
Valentias, 54 were in fact converted
Victorias, but the whole point of the
name change (Victory, Vindhya and
Vancouver were all considered) was
most likely to draw attention (on
safety grounds) to the different
loading capacities. Numerous types
have been dubbed the *Flying Pig* on
account of their handling or

appearance, but the Valentia must be
unique in that the name also referred
to the smell of the thing; when it
served as a w/op trainer with No. 1
Signals School, so many meals were
regurgitated inside the lumbering,
lurching beast that the pilots' open
cockpit came to be regarded as an
essential safety feature. A commercial
offshoot of the Victoria/Valentia line,
the one-off Vanguard (G-EBCP),
served only briefly with Imperial
Airways before it crashed at
Shepperton on 16 May 1929.

Viceroy
Name originally intended for the
famous Vickers Viscount airliner, but
later deemed inappropriate when
India became independent in 1947.
Powered by four Rolls-Royce Dart
turboprops (or propeller-turbines, as
they were then usually known), the
short-bodied Viscount 630 prototype
G-AHRF was loaned to BEA to fly
the world's first turbine-powered
commercial flight (Northolt to Le
Bourget) on 29 July 1950. BEA later
christened its own Viscount fleet the
Discovery class.

Vicky Ten
RAF-speak for the beautiful BAC
(Vickers) VC 10, along with the *Big
White Bird*, and in more recent years
Skoda as the original C1 transports
have begun to show their age. Built in
commercial Standard and Super
models (of which the original new-
build RAF aircraft were hybrids,
unlike the later tanker conversions)
there was also a little known project in
the mid sixties for the Superb, with
double-deck seating for up to 265.

Vigie
Diminutive form of **Vigilante**, the

handsome North American A-5 (ex-A3J) twin-jet attack bomber of the US Navy, which perhaps found its true niche in RA-5C reconnaissance guise. An attack version was also reportedly offered to the USAF in 1957 as the Retaliator.

Vigilante
Although agricultural aircraft or 'crop-sprayers' are often perceived as dull, uninteresting types, many in fact are quite feisty little machines, as befits the need to lift a heavy load and drop it accurately whilst avoiding hazards like power lines etc. The prime illustration of this was perhaps the one-off Ayres V-1-A Vigilante of 1989, a low-cost two-seat, close-support or surveillance version of the Turbo-Thrush ag-plane. Armament could be carried on eleven NATO-standard external hardpoints, the cockpit was armour-plated, and an endurance of no less than seven hours was claimed. The **Vigilante** was derived from the Turbo-Thrush S2R-T65/400 NEDS (Narcotic Eradication Delivery System) aircraft, nine of which were purchased by the US State Department between 1983 and 1985 and used to destroy drug crops in Colombia, Thailand, Guatemala, etc. Ayres Corp. purchased the rights to the basic Thrush ag-plane from Rockwell (to whom it was known as the Thrush Commander) in November 1977, and has also developed the Bull Thrush with smaller wings, enlarged chemical hopper, and a more powerful Wright R-1820 Cyclone radial engine in place of the standard P&W R-1340 Wasp. The Turbo Sea Thrush is a firefighting floatplane model developed by Canadian company Terr-Mar, and uses a probe to fill the hopper as the aircraft taxies over a lake or river.

Viper
Seldom has any aircraft been so important yet so badly misnamed as the General Dynamics (now Lockheed) F-16, officially the Fighting Falcon. The nose, intake and tail all scream **Shark**, *Flight* magazine proposed Mustang II after the classic P-51, and **Condor** was widely reported to have been chosen in the summer of 1979, only to be withdrawn; and if the USAF wanted another bird of prey to go with the F-15 Eagle, did it really need that awful 'Fighting' prefix to set it apart from the Dassault Falcon corporate jet? Small wonder then that pilots throughout America and Europe have adopted their own name of *Viper*, although the title is comparatively scarce outside of the immediate F-16 community. Among lesser nicknames are *Lawn Dart* after the shape, and of course the early *Electric Jet* referred to the pioneering Fly-by-Wire (FBW) flight control system; a variation on the name was *Electric Wedge* for the cranked-arrow wing F-16XL (alias F-16E and F) flown in prototype form in the mid-eighties, but passed over in favour of the F-15E. The Agile Falcon was a proposed development with an enlarged version of the standard cropped-delta wing, and Night Falcon was the title used to highlight the Block 40 production F-16C, compatible with Martin Marietta LANTIRN pods (Low-Altitude Navigation and Targeting Infra-Red, Night).

Voila
Widely used nickname within the Imperial Russian Flying Corps and (from 1917) Soviet Air Force for the French-designed Voisin LA pusher biplane.

Volksjager
Propaganda it may have been, but **Volksjager** or 'People's Fighter' perfectly summed up the extraordinary concept behind the Heinkel He 162, a simple single-engined jet fighter intended to be built in the thousands by mainly semi-skilled labour, and flown by the Hitler Youth after training on nothing more than gliders. Known unofficially by Heinkel as the Spatz (Sparrow), the He 162 first flew on 6 December 1944, just 90 days after the release of the specification, and though at least 170 had been delivered by VE Day it is doubtful whether any ever saw combat. The name of Salamander, so often used for the He 162 itself, was actually the name of the development and production programme.

Wackett Wonder
Although the advanced twin-engined attack prototypes built in Australia in WW2 were formally known as the CA-4 Wackett Bomber and CA-11 Woomera, many preferred *Wackett Wonder*, in recognition of Commonwealth Aircraft's Wing Cdr. L.J. Wackett.

Wag-a-Bond
Replica of the late forties-vintage Piper PA-15 Vagabond two-seat lightplane, available since 1978 in either kit or plan form from Wag Aero Inc. of Wisconsin. Two sub-versions are available, the Classic reproducing the basic PA-15, and the Traveler being a camper version with an extra door, increased area convertible for sleeping purposes, and extra baggage space.

Wakazakura
Wakazakura (Young Cherry) was the title of an unpowered recoverable

training version of the Japanese Yokosuka MXY-7 Ohka (Cherry Blossom) kamikaze suicide aircraft. Unlike most other kamikaze ('Divine Wind') aircraft, the tiny rocket-powered Ohka (Allied reporting name 'Baka') was purpose-designed for the role, and the only operational model, the Ohka 11, was carried into the attack slung underneath a Mitsubishi G4M bomber (Allied reporting name 'Betty').

Warferry
Name applied to the eight Foster Wickner Wicko GM1 cabin monoplanes impressed into the RAF in 1940.

Warrior
An armed version of Italy's stylish Agusta (formerly SIAI-Marchetti) SF-260 piston/turboprop trainer, the SF-260W Warrior first flew in May 1972 and was also available in SF-260SW Sea Warrior form for surveillance and search-and-rescue duties. In the sixties the civilian version was marketed in the US by Waco as the **Meteor**, and Genet is the name used by the Rhodesian/Zimbabwean Air Force for both the SF-260C and W models.

Warthog
To those who object to the re-use of famous names from the past, the greatest travesty of all was the Thunderbolt II title announced in April 1978 for the Fairchild A-10 close-support aircraft, replacing the provisional name of Sparrowhawk. The name was rarely used however, except to be punned in the early days as *Thunderblot* or just *Blot*, but a British magazine's assertion that it would 'inevitably' become another *Jug* like the WW2 P-47 Thunderbolt fighter proved well wide of the mark.

144

Those with a more developed sense of tradition had already christened the 'Ugly but Well-Hung' A-10 the *Warthog* after earlier Republic *Hogs* (Fairchild having taken over Republic in 1965), and by the time of the 1991 Gulf War the name had become so commonplace that many media reports evidently mistook it for a formal title. Lesser nicknames include *Porker* and *SLAT* (Slow, Low, Aerial Target). The last of 713 A-10s was completed in 1984.

Flying Washboard

American term for various aircraft with corrugated-metal skins (for strength) in the twenties and thirties, notably the Ford Tri-Motors and the long line of Junkers transports.

Washington

RAF name for the 88 Boeing B-29 Superfortress bombers loaned by the USA between 1950 and 1958.

Wet Wot

Designed pre-war by R.J. Currie, the single-seat Wot light sporting biplane was reputedly named after the frequent utterance of the partially deaf joiner employed on its construction. The two original Wots were both destroyed in an air-raid on Lympne (Kent) in 1940, but from 1958 a further fourteen were built by various individuals and organisations; variations on the theme included the Hot Wot (60hp Walter Mikron II engine), Hotter Wot (65hp Mikron III), Jet Wot (Rover IS 60 turbine) and of course the *Wet Wot* with an unsuccessful twin-float arrangement. In 1967 six Wots were built by Slingsby Aircraft to resemble Royal Aircraft Factory SE5as for film work, cunningly reproducing the WW1 scout's long exhaust tubes with lengths of plastic drainpipe.

Flying Whale

Somewhat contrived Soviet 'nickname' for the Antonov An-8 (NATO – Camp) twin-turboprop military freighter first revealed in the 1956 Tushino flypast.

Whale (i)

Popular but slightly misleading nickname for the German LFG ('Roland') CII two-seater of 1916/17, for despite its deep and capacious fuselage the Walfisch was remarkably streamlined for its time.

Whale (ii)

Largest and heaviest type ever to operate routinely from aircraft carriers, the Douglas A-3 twin-jet attack bomber was known on US Navy decks almost exclusively as the *Whale*, the formal title of Skywarrior being strictly for outsiders and FNGs (... New Guys). *Electric Whale* and *Queer Whale* were names for both the A3D-1Q and -2Q electronic warfare models (cf *Q-Bird*) and the awkward-looking approach to a carrier deck was known as the 'whale dance'. With an emergency 'chute instead of ejection seats, cynics claimed the original A3D designation really stood for 'All 3 Dead'. The A3D-1 first joined the fleet in 1956, and eventually equipped a total of thirteen Heavy Attack squadrons.

Whispering Death (i)

Widely quoted Japanese term for the Bristol Beaufighter, better known to its friends simply as the *Beau*. Someone once described it as 'two bloody great engines hotly pursued by an aeroplane', and in fact at one stage it had something of a fearsome reputation in the RAF, particularly for certain landing characteristics. This was reputedly dispelled at one station when an unexpected arrival put on an effortless textbook approach and three-point landing, and the 'Ace' that was met from the Beaufighter by a small crowd of admirers turned out to be a small and bemused female ATA ferry pilot. As with the Spitfire, Typhoon and Whirlwind, the Beaufighter name was sometimes adapted to suit its role and armament, hence rocket-toting *Rockbeaus* or *Flakbeaus* and torpedo-carrying *Torbeaus*.

Whispering Death (ii)
Vietcong term for the US Army's Grumman OV-1 Mohawk twin-turboprop battlefield recce aircraft. Machines with the long Side-Looking Airborne Radar (SLAR) pod slung under the fuselage (principally OV-1Bs and Ds) were considered by their crews to be the male of the species. The most famous of all Mohawks was the legendary 'Ole Yeller', so named because of all the zinc chromate paint over its battle damage repairs; finally abandoned by its crew because of an undercarriage malfunction, the veteran of over 500 missions declined to be buried at sea and instead turned and made two overflights of its Vung Tau base before obligingly crashing in the scrapyard.

Whispering Giant
One of the few promotional or advertising names to gain any degree of popular usage was this title used by BOAC for the Bristol 175 Britannia, the beautiful long-range turboprop that once seemed destined for greatness. The Aeronautical Engineers Association suggested calling it the Empire class in 1953 (with Empire Traveller as the flagship) but as with the **Skyliner** proposal the name was not taken up.

Partly as a result of its protracted gestation and development problems with the Bristol Proteus engines, just 85 were built in the UK, including twenty for the RAF.

Canadair developments included the RCAF's mighty CP-107 Argus (Canadair CL-28) maritime aircraft with a new shorter fuselage, twin 18ft weapons bays, and Wright Turbo-compound engines giving a normal endurance of 26½ hours; 33 were built between 1957 and 1960. The RCAF also instigated a transport development of the Argus with Rolls-Royce Tyne engines and a 12ft stretch over the Britannia, known in RCAF service as the CC-106 Yukon (twelve aircraft) and in civil swing-tail freighter form as the CL-44D, sometimes styled 'Forty-Four'. Commercial production came to 27, including four CL-44Js (alias Canadair 400) with a further 15ft stretch, operated as high-density, 189-seat airliners by the Icelandic carrier Loftleidir; Canadair also offered a short-range 214-seater variant in the mid-sixties as the Candairbus, but there were no takers. More radical still is the one-off CL-44O Airlifter converted from a CL-44D in 1969 by Conroy Aircraft, with a greatly expanded upper fuselage lobe of 15,000 cubic ft capacity; this long-serving machine (still in service with Texas-based Buffalo Airways in 1994) was for many years known by its own, original nickname of *Skymonster*, but in more recent times this has unfortunately given way to *Guppy*, after Jack Conroy's better known Boeing Stratocruiser (cf **Strat**) conversions.

Whistling Turtle
Belgian Air Force nickname for the well-known Fouga CM 170 Magister

French jet trainer with its squat sit on the ground and distinctive note from its twin Marbore engines. Also known as the *Dinky Toy*, the V-tailed Magister was famous for many years as the mount of the Patrouille de France aerobatic team, and also served with the Aeronavale as the Zephyr (formerly the Magister Marine and Esquif). The unbuilt, swept-wing CM 195 model was known somewhat optimistically as the Mach 1, and the CM 171 Makalu of 1956 was a one-off test-bed for the Turbomeca Gabizo turbojet, featuring enlarged nacelles separated from the fuselage by a short, new wing centre-section.

Under the AMIT (Advanced Multi-Mission Improved Trainer) programme of the early eighties, Israel Aircraft Industries updated 86 Heyl Ha'Avir Magisters with uprated engines and cockpit as the Tzukit (Thrush). Aerospatiale (successor to the old Fouga company) had themselves begun studying an updated derivative in the mid-seventies as the Fan Jet 600, which took its title from the projected DF600 turbofans; the project eventually crystallised in 1978 with the first flight of the private-venture Fouga 90 (one only, F-WZJB) with Astafan engines and staggered seating, but the type failed to sell and was abandoned in 1980.

Whistling Wheelbarrow
Well-known nickname for the Armstrong Whitworth AW 650/660 Argosy freighter, with its twin tailbooms and four RR Dart turboprops. Last aircraft to bear the famous AW name (and their second Argosy) the type was known in the project phase as the Freightliner, and unbuilt developments included the AW 670 Air Ferry with a double-deck fuselage for six to eight cars and 30

passengers, and the AW 671 Airbus with a similar fuselage seating up to 126 passengers. The name Commuter was at one stage associated with a projected stretched 108-seat variant of the production Argosy 200 series freighter.

The RAF's Argosy C1 never quite shook off an image of being an unsuitable civil type dumped on the Service (56 out of a total of 73 Argosies were built for the RAF) and a popular jibe was that the most it could safely carry was a verbal message. The new thimble-nose radome gave rise to *Flying Tit* or *Whistling Tit*, whilst camouflaged examples conjured up memories of WW2 heavies with their twin tails and four props, and were sometimes referred to as *Brown Bombers*.

Whitebait
Another Royal Flying Corps fish nickname, this time for the pre-WW1 Breguet L1, L2 and G3 tractor biplane series, a fragile-looking contraption with an all-flying cruciform tail mounted on a universal joint, and single-strut interplane bracing in place of the more usual paired arrangement. Aluminium cladding on the forward fuselage earned the French machine the nickname *Tin Whistle* in the Royal Naval Air Service.

White Rocket
USAF nickname for the supersonic Northrop T-38 Talon, the pure trainer counterpart to the F-5 fighter series.

White Tail
Industry term for an airliner built 'on spec' but as yet unsold, i.e. lacking any operator's logo or motif.

Widow Maker
In view of the terror it evoked in many

of its crews during training, perhaps the Glenn L. Martin Co. should have stayed with their original title of Martian for the B-26 Marauder twin-engined medium bomber. A vicious combination of high wing-loading and propeller malfunctions quickly earned the B-26 such grim epithets as the *Baltimore Whore* and *Flying Prostitute* (because of its small wing area they said it had no visible means of support), *Murderer*, *Killer* and *B-Dash-Crash*, but of course the one that really stuck was *Widow Maker*. *Flying Torpedo* and *Bumble Bee* (from the noisy engines) are mild in comparison, but in fact once the bugs were ironed-out and the training was improved, the accident rate quickly tailed off, and the B-26 ended the war in Europe with the lowest operational loss rate of any USAAF type.

Wild Weasel
Like so many Vietnam-era names, **Wild Weasel** originated as a programme title (Ferret and Mongoose were also considered) but in time came to be an accepted name for the actual aircraft adapted for the USAF's SAM (Surface-to-Air Missile) lethal suppression role. The original Wild Weasel 1 was a conversion of the North American F-100F Super Sabre, seven of which were urgently fitted with specialised avionics and weaponry for locating and destroying SAM radar sites; the actual mission was given the codename Iron Hand, and a strike package usually consisted of a single Weasel and four Republic F-105 Thunderchief strike aircraft.

The Wild Weasel II was a conversion of the McDonnell F-4C Phantom with essentially the same avionics hung in pods under the wings, hence the nickname *Weasel in a Can*, but because of inadequate systems-

integration it was not employed operationally. The first really mature models, entering service in Vietnam in spring 1966, were the F-105F (86 conversions of the standard F-105F two-seater) and F-105G (60 converted Fs), both of which were entitled Wild Weasel III. Three North American T-39 Sabreliners converted to train Electronic Warfare Officers (EWOs) for F-105 Weasels were designated T-39Fs and nicknamed *Teeny Weeny Weasels* at Nellis AFB; EWOs on Weasels are known colloquially as 'Bears' (the back seat is also the 'Pit') from the way they are toted around by the pilot like a tame bear on a chain. The Wild Weasel IV was another F-4C adaptation which entered service in 1969, but never fully replaced the F-105 versions.

The current F-4G Advanced Wild Weasel is not only the last Phantom version in US front-line service, but could also be the last true dedicated Weasel, although the F-15 Eagle is expected to acquire a defence-suppression capability around the turn of the century, using the AGM-88 HARM (high-speed anti-radiation missile). A total of 134 F-4Gs were converted from standard F-4Es first entering service in 1977, and the type has also been operated in Hunter-Killer pairs with the GD/Lockheed F-16. In such a role the F-16 is sometimes loosely referred to as a **Wild Weasel**, but even the latest Block 50D F-16C with the HARM Targeting System reportedly has, at most, just 80 per cent of the F-4G's target-detection capability. Since the Gulf War the **Wild Weasel** name has also become somewhat bastardised by its indiscriminate application to electronic warfare types like the GD/Grumman EF-111A Raven and even the US Navy's Grumman

EA-6B Prowler. Unlike the true
USAF Weasels, both were initially
unarmed and thus unable to actually
attack SAM radars, although with the
introduction of the ADVCAP
(Advanced Capability) EA-6B in the
early nineties the Prowler acquired the
capability to launch HARM missiles.

Willie the Whale
Not to be confused with the same
company's A-3 *Whale*, this was the
almost universal nickname of the
fifties-vintage Douglas F3D Skyknight
naval night-fighter jet, which also
soldiered on well into the Vietnam
War as the EF-10B electronic warfare
aircraft. In later years the portly F3D/
EF-10B was known rather unkindly as
the *Drut*, which is a rare example of
an ananym i.e. a word that should be
read backwards.

Wimpy
Whilst there are those who referred to
the RAF's famous Vickers Wellington
as the *Basketweave Bomber* (after the
geodetic construction) or the *Flying
Cigar*, all other nicknames pale into
insignificance beside *Wimpy*, after
Popeye's docile, hamburger-eating
sidekick J. Wellington Wimpy. As the
Wellington established itself as the
backbone of Bomber Command in the
early war years, the affectionate title
attained a currency with servicemen
and public alike that is probably
unequalled among non-diminutive
type nicknames. The Wellington name
itself was announced by the Air
Ministry in September 1936, replacing
Crècy which Vickers had been using
for the B9/32 since the previous June.
Wedding Ring Wimpys were those
fitted with a 48ft diameter magnetic
coil for minesweeping duties, Coastal
Command aircraft were known as
Goofingtons (from 'goofing around')

or *Fishingtons* when carrying 'Tin
Fish' (torpedoes), and the Mk VIII
and XIII with a row of four ASV (Air-
to-Surface Vessel) radar masts
projecting vertically from the spine
were known as *Sticklebacks* after the
pond fish. A trio of Wellingtons
(Z8570, W5389 and W5518) were used
as jet engine testbeds from 1943 and
were known to Vickers by the
codename Squirter.

Windicator
The Royal Navy received 50 Vought
Sikorsky V-156 scout bombers (the
residue of a French contract) from
America in 1941, and in fact the
British name of Chesapeake was the
first to be given to the type. The two-
seater monoplane had been in service
with the US Navy as the SB2U since
1937, but it was not until the final
SB2U-3 version that the name
Vindicator (punned variously as
Windicator and *Wind Indicator*) was
introduced and applied retroactively
to all domestic models.

Winged Can-Opener, The
In some ways analogous to the
Fairchild A-10 with its twin engines,
heavy-calibre gun (on some models)
and armoured cockpit, the Henschel
Hs 129 close-support aircraft was,
however, not a popular type with
Luftwaffe pilots; many compared it to
the same company's railway locos and
tanks, dubbing the cramped cockpit
the 'Panzerkabine', and the poor
handling was exacerbated by an
unusually short control column which
shared its nickname with an intimate
part of the male anatomy ...

Wobblin' Goblin
For years the American Department
of Defense steadfastly denied the
existence of any so-called Stealth

149

Fighter, and the embarrassment they supposedly felt over the famous model kits of the sleek Lockheed F-19 only seemed to confirm both its existence and its appearance. In fact the whole affair was a masterpiece of disinformation, for when the first picture of the actual Lockheed F-117 was released on 10 November 1988 it revealed a 'faceted' arrowhead machine bearing virtually no relation to the smooth shape that had seemed such an open secret. The F-19 designation has still not been accounted for though, and both the F-117 and Stealth Fighter labels are misnomers, as the aircraft is in no real sense a fighter, being used instead (as in the Gulf War) for precision strikes on high-value targets. Furthermore, F-117 would appear to belong to the *Century Series* of designations which supposedly ended in 1962 with the F-111; the most likely explanation is that the intervening designations (i.e. F-112 to F-116) were applied to Soviet types secretly evaluated in the USA. In time the F-117 became known more or less formally as the Nighthawk after the unit title of the 37th TFW which once operated it, although its pilots usually refer to it undramatically as the *Black Jet*. Other nicknames have included *Beast* and of course the infamous *Wobblin' Goblin* sobriquet which so upset Lockheed, referring to an early controllability problem; *Ghost* and *Specter*, whilst clearly influencing the *Goblin* name, belong firmly to the days of the imaginary F-19. Senior Trend was the name of the Black programme under which the F-117 was developed by the Lockheed 'Skunk Works', whilst Have Blue was the programme title for the preceding XST proof-of-concept aircraft, broadly similar in appearance to the F-117 but only about half its size; two

XSTs were built, and both crashed on test flights.

Wokka Wokka
Echoic RAF name for the Boeing Chinook helicopter; the name is a double version of the more general slang for helicopters (cf *Chopper*) denoting the fact that the Chinook is the RAF's only twin-rotor helicopter. In Vietnam the powerful, carry-anything CH-47 Chinook was sometimes crudely referred to as the *Shithook*, and the ACH-47A *Go-Go Bird* was an experimental heavy gunship model with 2,000lb of armour plating, a grenade launcher in the nose, and various combinations of 20mm Gatling guns, 7.62mm machine-guns and 2.75in rockets; four were evaluated in Vietnam, two were lost in action and the project discontinued.

Wop
Best known simply as the *Wop*, the RAF's inter-war Westland Wapiti general-purpose biplane was also dubbed *The Rock* by wags who said it was as steady as the Rock of Gibraltar – although only half as fast. The Wallace, a major derivative first flown in 1931 as the PV6 or Wapiti VII, was known as the *Nellie* after the music hall artiste Nellie Wallace.

Wren
Name formerly intended for the Swiss FFA AS202 Bravo trainers used by the British Aerospace College, Prestwick.

X-Avia
Stillborn, late seventies project for a Westernised version of the Yak-40 (NATO – Codling) tri-jet feederliner, to have been powered by Garrett AiResearch TFE-731-3 turbofans and built by ICX Aviation at Youngstown, Ohio using the original Soviet jigs.

X-Boat
Dutch Navy name, taken from the
prefix to their serials, for the Dornier
Do 24K tri-motor flying boats
operated in the East Indies; a dozen
Do 24K-1s were delivered from
Dornier's Swiss subsidiary, but licence
production of 48 Do 24K-2s in
Holland by Aviolanda was halted by
the German invasion with just 25
delivered.

X-Engine Airplane
Lockheed nickname for the
experimental XP-58 Chain Lightning,
an enlarged development of the
famous P-38 Lightning planned
around no less than six different types
of engine before it finally flew on 6
June 1944 with 24-cylinder Allison V-
3420s.

Yellow Peril
The formerly traditional use of yellow
paint schemes on British,
Commonwealth and American
trainers, combined with the inevitable
accidents, has meant that innumerable
types have been branded with this
epithet at some stage; the Boeing
Stearman Kaydet, Cessna Crane/
Bobcat, N.A. Harvard/Texan, Miles
Magister, even the DH Tiger Moth.
Clockwork Mouse has been associated
with training aircraft since at least the
WW1 Airco DH6, when it was used by
the 'Ack Emmas' (Air Mechanics) to
describe the low-powered aircraft
making continual circuits and bumps.
Since WW2 *T-Bird* has found favour
in the US, whilst *Tub* relates to a two-
seat conversion trainer model of what
is normally a single-seater, for
example the BAC Lightning T4/5 and
Lockheed F-16B; the Soviet/CIS
counterpart is *Sparka* (*Double*). *Twin-
Sticker* is used to describe a dual-
control version of an operational

two-seater, i.e. the RAF's Tornado
GR 1(T).

Zero
The most famous Japanese aircraft of
all time is undoubtedly the Mitsubishi
Zero fighter, which in the early stages
of the Pacific air war had the same
almost-invincible reputation that the
Japanese soldier had on the ground.
The hitherto complacent Western
attitude to Japanese aircraft was
rudely shattered when the **Zero** was
first encountered at Pearl Harbor on
7 December 1941, and salt was later
rubbed into the wound when
American newspapers began carrying
sensational (and false) stories that it
was a re-hash of either the Northrop
3A which disappeared mysteriously
off the Californian coast in July 1935,
or its derivative the Vought V-143,
which was sold to Japan two years
later. In nomenclature terms the *Zero*
nickname was unusual in being widely
used (albeit in different forms) by
both sides, and had its origin in the
long-standing Japanese practice of
designating aircraft from the last two
digits of the year in which they were
accepted for military service. In the
Japanese calendar 1940 was the year
2600, but unbeknown to the West this
also coincided with a change to using
only the last digit; thus in the West the
new fighter was thought to be the
'Type 00' and was quickly nicknamed
Zero, whilst in its homeland it was
officially the Navy Type 0 and
nicknamed *Reisen* (from 'Rei Shiko
Sentoki' or 'Type 0 Fighter'). More
formally the type had the alternative
Navy designation of A6M and the
Allied Reporting name of 'Zeke'.

Zerstorer
Although the German word for
Destroyer was strictly applied to a

ZERSTORER

class of aircraft rather than any
specific type, it became almost as
closely connected with the twin-
engined Messerschmitt Bf 110 as the
contemporary term *Stuka* was with
the Ju 87. Descended from the WW1
Battle Planes, the concept was for a
heavily-armed, long-range bomber
escort, but during the Battle of Britain
the once elite Zerstorergruppen
suffered the indignity of themselves
having to be escorted by single-
engined Bf 109s.

APPENDIX 1

Allied Reporting Names for WW2 Japanese Aircraft

With the start of the war in the Pacific, Allied intelligence was found badly wanting in respect of Japanese aircraft. The problem was compounded by the differing and complex (at least to a Western mind) designation systems used by the Japanese Army and Navy, and so to rationalise available information a series of reporting names was evolved in mid-1942 by a unit based in Melbourne known as the Directorate of Intelligence, Allied Air Forces, SW Pacific Area. Initially, Fighters and Floatplanes were given masculine Christian names whilst the remainder received feminine names; in time, however, the system was expanded to encompass bird names for gliders, tree names for trainers, and transports were given names beginning with the letter T. Many name allocations reflected private (or not so private) jokes concerning the staff responsible for the system, their friends, families, etc.; for example, the Mitsubishi G4M bomber became the Betty because its rounded gun cupolas reminded someone of a certain buxom American nurse. Inevitably there were anomalies, notably with types being allocated more than one name, and with foreign types that failed to appear wearing the infamous Japanese 'meatball'. But overall the system worked well, and to this day remains the most manageable form of identification for WW2 Japanese types.

As well as the Allied reporting names, the tables give a brief description of the aircraft concerned, and formal Japanese nomenclature as follows:

Short Titles: The Army and Navy each had their own short titles which originated in the early days of an aircraft project and usually stayed with it throughout its career, in parallel with the Type Number (see below) allocated upon entry into service. The Army system was based on Ki (Kitai or Airframe) or Ku (from the verb Kakku, 'to glide') numbers, whilst the Navy method was basically similar to the USN system, comprising a role letter, manufacturer's number within that role, manufacturer or designer's letter, and model number. Thus A6M2 for the Zero denoted that it was a fighter (A), 6th fighter type from Mitsubishi (6M), second model (2). Later in the war a simplified system of names and model numbers was adopted, e.g. Suisei Model 33; when known, these model numbers were suffixed to the Allied name, e.g. the Suisei Model 33 became Judy 33.

Japanese Names: Japanese aircraft names fell into the following categories: Blossoms (kamikaze suicide aircraft), Clouds (Seaplanes and Recce types), Light (Night-Fighters), Mountains (Attack Bombers), Seas (Maritime Patrol aircraft), Skies (Transports), Stars (Bombers), Thunder and Lightning (Land-based fighters), Trees and Grasses (Trainers) and Wind (Carrier-borne and Seaplane Fighters).

APPENDIX 1

Army/Navy Type Numbers: Basically a formal description of the aircraft's role, prefixed by the Japanese calendar year (660 ahead of the West) in which it was accepted for service. For example, the well-known Mitsubishi Navy Type 96 Attack Bomber (Nell) was accepted in the Japanese year 2596 (1936 in the West). Where two types were accepted for the same role in the same year, a suffix would be added thus: Type 96-1, Type 96-2, etc. Navy designations effectively started again from scratch in the Japanese year 2600 (leading to the famous Type 0 or Zero) whilst the Army initially followed its various Type 99s (from the year 2599) with Type 100s.

Abdul	Name allocated to expected development of **Claude** with retractable undercarriage, which did not materialise.
Adam	Allocated to mythical Nakajima floatplane.
Alf	Kawanishi E7K, Navy Type 94 reconnaissance seaplane. Twin-float biplane, often used as catapult-launched ship's scout.
Ann	Mitsubishi Ki-30, Army Type 97 light bomber. Radial engine, fixed spatted undercarriage. Some used by Thais in 1940 conflict with French; known as Nagoya from location of Mitsubishi factory.
Babs	Mitsubishi C5M, Army Type 97 Command reconnaissance plane, Ki-15. Single-engined, fixed-gear monoplane.
Baka (Fool)	Yokosuka MXY-7 Ohka (Cherry Blossom) Navy special attacker. Air-launched, rocket-powered kamikaze aircraft.
Belle	Kawanishi H3K, Navy Type 90-2 flying boat. Biplane flying boat based on Short Calcutta. Five only, not used in WW2.
Ben	'Nagoya Type 00'. Duplicate name originally applied to **Zeke** (Zero). Nagoya name, taken from aircraft shot down at Pearl Harbor was thought to be name of manufacturer (cf **Ann**).
Bess	German Heinkel He 111 bomber erroneously thought to be built under licence by Aichi or Nakajima.
Betty	Mitsubishi G4M, Navy Type 1 attack bomber. Famous twin-engined medium bomber.
Bob	Nakajima E2N, Navy Type 15 reconnaissance seaplane. Twin-float biplane, withdrawn from front-line use by WW2.
Buzzard	Kokusai Ku-7 Army experimental transport glider. Twin-boom, tank-carrying glider.
Cedar	Tachikawa Ki-17, Army Type 95-3 primary trainer. Tandem-seat training biplane.
Cherry	Yokosuka H5Y, Navy Type 99 flying boat. Twin-engined, parasol-wing flying boat.
Clara	Tachikawa Ki-70, Army experimental command reconnaissance plane. Twin-engined monoplane.
Claude	Mitsubishi A5M, Navy Type 96 carrier fighter. Well-known open-cockpit, fixed-gear fighter.
Clint	Duplicate name originally applied to **Nate**.
Cypress	Kokusai Ki-86, Army Type 4 primary trainer. Kyushu (Watanabe) K9W, Navy Type 2 primary trainer, Momiji (Maple). German Bucker Bu 131 Jungmann trainer biplane built under licence.

154

Dave	Nakajima E8N, Navy Type 95 reconnaissance seaplane. Radial-engined biplane, single main float.
Dick	Seversky A8V, Navy Type S fighter. Name applied to projected Japanese copy of Seversky 2PA two-seat fighter monoplane, twenty of which were supplied from the US in 1938.
Dinah	Mitsubishi Ki-46, Army Type 100 command reconnaissance plane. Well-known high-speed streamlined twin.
Doc	Projected licence-built Messerschmitt Bf 110. Did not exist.
Doris	Unidentified twin-engined bomber. Initially regarded as a newly identified type, but may have been **Lily**.
Dot	Duplicate name originally applied to **Judy**.
Edna	Mansyu Ki-71, Army experimental tactical reconnaissance plane. Higher-powered, retractable undercarriage version of **Sonia**. No production.
Emily	Kawanishi H8K, Navy Type 2 flying boat. Famous four-engined flying boat, broadly similar in appearance to Short Sunderland.
Eve	Civil version of **Louise** (the Ohtori or Phoenix) mistaken for a new bomber type.
Frances	Yokosuka P1Y Navy bomber Ginga (Milky Way), P1Y1-S Navy night fighter Byakko (White Light), P1Y2-S. Navy night fighter Kyokko (Aurora). Initially given masculine name of Francis when original bomber version was thought to be a twin-engined heavy fighter.
Frank	Nakajima Ki-84, Army Type 4 fighter, Hayate (Gale). Highly capable, single-seat fighter, first flown April 1943.
Fred	Name allocated to expected licence-built Focke-Wulf Fw 190 German fighter.
Gander	Kokusai Ku-8, Army Type 4 special transport glider. Twenty-seat assault glider.
George	Kawanishi N1K, Navy interceptor fighter Shiden (Violet Lightning). Powerful landplane derivative of **Rex**.
Glen	Yokosuka E14Y, Navy Type 0 light reconnaissance seaplane. Small two-seater launched from submarines.
Goose	Original name applied to **Gander**, changed to avoid confusion with Grumman JRF Goose.
Grace	Aichi B7A, Navy carrier attack bomber Ryusei (Shooting Star). Heavy, single-engined attack aircraft.
Gus	Nakajima AT-27. Unbuilt fighter project with contra-rotating props and twin engines fore-and-aft of cockpit.
Gwen	Duplicate name originally applied to a version of **Sally**.
Hamp/Hap	Initially thought to be a new fighter and named Hap after General 'Hap' Arnold; latter did not approve and name changed to Hamp, supposedly to avoid confusion with the word 'Jap'. Aircraft later realised to be a clipped-wing variant of **Zeke** and renamed Zeke 32.
Hank	Aichi E10A, Navy Type 96 reconnaissance seaplane. Single pusher-engined biplane, similar in layout to Supermarine Walrus.
Harry	Mitsubishi TK-4. Unbuilt push-pull, twin-engined fighter project.
Helen	Nakajima Ki-49 Donryu, Army Type 100 heavy bomber. Well-

APPENDIX 1

armed bomber, but disappointing performance. Donryu was name
of a shrine at Ota, and translates as 'Storm Dragon' or, more
informally as 'Dragon Swallower'.

Hickory   Tachikawa Ki-54, Army Type 1 trainer/transport. Multi-purpose,
           twin-engined, trainer-comms aircraft.
Ida   Tachikawa Ki-36, Army Type 98 direct cooperation plane. Fixed-
           gear monoplane, later adapted to training role as Ki-55, Army Type
           99 advanced trainer.
Ione   Aichi AI-104. Unbuilt, tri-motor seaplane project.
Irene   Name allocated to expected licence-built Junkers Ju 87 Stuka.
Irving   Nakajima J1N. Initially intended as a heavy fighter (hence
           masculine reporting name) but later converted as Navy Type 2 land-
           based reconnaissance plane, and highly successsful Navy Type 2
           night-fighter Gekko (Moonlight).
Jack   Mitsubishi J2M, Navy interceptor fighter Raiden (Thunderbolt).
           Portly but powerful fighter which underwent protracted
           development.
Jake   Aichi E13A, Navy Type 0 reconnaissance seaplane. Twin-float
           scout, radial engine.
Jane   Duplicate name originally applied to **Sally**.
Janice   Name allocated to expected licence-built Junkers Ju 88.
Jean   Yokosuka B4Y, Navy Type 96 carrier attack bomber. Single-
           engined biplane, obsolete by WW2.
Jerry   Heinkel He 112B-0 fighter monoplane, twelve of which were
           imported in 1938 (eighteen more cancelled) and designated A7He1
           Type He Air Defence fighter.
Jill   Nakajima B6N, Navy carrier attack bomber Tenzan (Heavenly
           Mountain). Three-seat, radial-engined monoplane.
Jim   Duplicate name originally allocated to **Oscar.**
Joan   Name allocated to four-engined flying boat later found not to have
           existed; type reported may have been **Mavis**.
Joe   Fictional 'TK-19' fighter monoplane appearing in pre-war
           periodicals and thought to be impending new Army type.
John   Duplicate name originally allocated to **Tojo.**
Joyce   Reported as new, twin-engined Nakajima light bomber, possibly
           resulting from another sighting of Ohtori (cf **Eve**).
Judy   Yokosuka D4Y, Navy carrier bomber Suisei (Comet). Built with
           both radial and inline engines.
Julia   Duplicate name originally allocated to **Lily**.
June   Name originally applied to **Jake** in the belief that it was a floatplane
           version of **Val** bomber (hence feminine name).
Kate   Nakajima B5N, Navy Type 97-1 carrier attack bomber. Radial-
           engined torpedo bomber, used in China and at Pearl Harbor (cf
           Mabel).
Laura   Aichi E11A, Navy Type 98 reconnaissance seaplane. Light, pusher-
           engined biplane, little used.
Lily   Kawasaki Ki-48, Army Type 99 twin-engined light bomber. Widely
           used despite mediocre performance.

156

Liz	Nakajima G5N, Navy attack bomber Shinzan (Mountain Recess). Experimental four-engined heavy bomber based on Douglas DC-4E (not to be confused with production DC-4 and C-54) airliner, the sole example of which was sold to Japan in 1939. Four only completed
Lorna	Kyushu Q1W, Navy patrol plane Tokai (Eastern Sea). Specialised anti-submarine/maritime patrol twin.
Louise	Mitsubishi Ki-2-II. Army Type 93-2 twin-engined light bomber. Early thirties design, retired long before WW2.
Luke	Mitsubishi J4M Experimental interceptor fighter. Twin-boom pusher project, cancelled before completion of prototype.
Mabel	Mitsubishi B5M, Navy Type 97-2 carrier attack bomber. Built to same requirement as **Kate**, and because it was so little used and bore a close resemblance, was renamed Kate 61.
Maisie	Name allocated to supposed new medium bomber, but aircraft reported may actually have been **Nick**.
Mary	Kawasaki Ki-32, Army Type 98 light bomber. Single-engined, fixed-gear monoplane, withdrawn soon after start of war.
Mavis	Kawanishi H6K, Navy Type 97 flying boat. Well-known, four-engined, parasol-wing flying boat.
Mike	Projected licence-built Messerschmitt Bf 109. Did not exist.
Millie	Name allocated to expected licence-built Vultee V-11GB single-engined American bomber. Did not exist.
Myrt	Nakajima C6N, Navy carrier reconnaissance plane Saiun (Painted Cloud). Outstanding high-altitude recce type, also adapted as a night-fighter.
Nate	Nakajima Ki-27, Army Type 97 fighter. Fixed-gear monoplane, often confused with Navy's **Claude** (cf **Clint**).
Nell	Mitsubishi G3M, Navy Type 96 attack bomber. Well-known, twin-engined, twin-tailed bomber.
Nick	Kawasaki Ki-45, Army Type 2 two-seat fighter Toryu (Dragon Slayer). Highly successful, twin-engined night-fighter/ground-attack type.
Norm	Kawanishi E15K, Navy Type 2 high-speed reconnaissance seaplane Shiun (Violet Cloud). Single main float, contra-rotating props. Not built in quantity.
Norma	Fictional type illustrated in pre-war Japanese periodicals, and assumed to be a forthcoming new Army light bomber.
Oak	Watanabe/Kyushu K10W, Navy Type 2 intermediate trainer. Derivative (with altered tail) of North American NA-16 fixed-gear model (precursor of T-6 Texan/Harvard).
Omar	'Sukukaze 20' (Cool Breeze) fictional fighter with tandem radial engines, seen in a pre-war Japanese periodical and thought to be a drawing of a forthcoming type.
Oscar	Nakajima Ki-43 Hayabusa (Peregrine Falcon) Army Type 1 fighter. Mainstay Army fighter, in production throughout entire Pacific war period.
Pat/Patsy	Tachikawa Ki-74 Army Experimental long-range bomber.

	Originally named **Pat** in expectation that Ki-74 was to be a fighter, changed to feminine **Patsy** when it emerged as a twin-engined pressurised bomber. Too late for operational use.
Paul	Aichi E16A, Navy reconnaissance seaplane Zuiun (Auspicious Cloud). Three-seater, used mainly as a dive-bomber.
Peggy	Mitsubishi Ki-67, Army Type 4 heavy bomber Hiryu (Flying Dragon). World-class, twin-engined bomber, also used as a Navy torpedo bomber and named Yasukuni after a Japanese shrine. Ki-109 was an experimental version with a 75mm cannon to intercept Boeing B-29 Superfortress.
Perry	Kawasaki Ki-10, Army Type 95 fighter. Obsolete biplane, used as a trainer in WW2.
Pete	Mitsubishi F1M, Navy Type 0 observation seaplane. Streamlined biplane also used for fighter and light bomber duties.
Pine	Mitsubishi K3M, Navy Type 90 trainer. Single-engined, high-wing training monoplane.
Randy	Kawasaki Ki-102a Army experimental high-altitude fighter, Ki-102b Army Type 4 assault plane. Heavily armed, fast twin, too late to make much impact.
Ray	Duplicate name originally allocated to **Zeke**.
Rex	Kawanishi N1K, Navy fighter seaplane Kyofu (Strong Wind). Purpose-built fighter seaplane; sired **George** landplane.
Rita	Nakajima G8N, Navy experimental attack bomber Renzan (Mountain Range). Four-engined heavy bomber with tricycle undercarriage. Four only completed.
Rob	Kawasaki Ki-64 Army experimental high-speed fighter. Twin engines fore-and-aft of cockpit, each driving nose-mounted props. Single prototype flown in December 1943, but suspended as too complex.
Rufe	Nakajima A6M2-N, Navy Type 2 fighter seaplane. Successful floatplane version of Mitsubishi **Zeke**.
Ruth	Army Type I medium bomber. Fiat BR 20 Cigogna (Stork) twin-engined bomber, 85 of which were imported in 1938. Not operational in WW2.
Sally	Mitsubishi Ki-21, Army Type 97 heavy bomber. Widely used twin-engined bomber. Hopelessly vulnerable in later stages of WW2.
Sam	Mitsubishi A7M, Navy experimental carrier fighter Reppu (Hurricane). Promising Zero replacement, too late for WW2.
Sandy	Duplicate name originally allocated to **Claude**.
Slim	Watanabe E9W, Navy Type 96 light reconnaissance seaplane. Twin-float biplane carried by submarines.
Sonia	Mitsubishi Ki-51, Army Type 99 assault plane. Widely used fixed-gear monoplane.
Spruce	Tachikawa Ki-9, Army type 95-1 intermediate trainer. Tandem-seat, radial-engined biplane
Stella	Kokusai Ki-76, Army Type 3 command liaison plane. High-wing STOL monoplane inspired by Fieseler Fi156 Storch.
Steve	Mitsubishi Ki-73, Army experimental fighter. Long-range escort-

APPENDIX 1

	fighter project with inline engine. Cancelled before completion of prototype.
Susie	Aichi D1A, Navy Type 96 carrier bomber. Two-seat biplane loosely based on Heinkel He 66. Used only as trainers in WW2.
Tabby	Showa L2D, Navy Type 0 transport. Japanese-built Douglas DC-3. Original licence granted February 1938.
Tess	Douglas DC-2, six of which were imported pre-war.
Thalia	Kawasaki Ki-56, Army Type 1 freight transport. Development of **Thelma** with 4ft 11in fuselage stretch, and thus equating roughly to Lockheed Model 18 Lodestar.
Thelma	Kawasaki Type LO transport. Lockheed Model 14 Super Electra. Thirty imported pre-war and a further 119 built in Japan (cf **Toby**).
Theresa	Kokusai Ki-59, Army Type 1 transport. High-wing, fixed-gear twin. Little-used.
Thora	Nakajima L1N, Navy Type 97 transport. Ki-34, Army Type 97 transport. Twin-engined, low-wing, eight-seat transport.
Tillie	Kusho H7Y Navy special flying boat. Twin-engined, gull-wing flying boat, designed pre-war for long-range recce over US territory. One only completed.
Tina	Mitsubishi L3Y, Navy Type 96 transport. Transport variant of **Nell** bomber.
Toby	Name originally allocated to Japanese commercial Lockheed Model 14s (cf **Thelma**).
Tojo	Nakajima Ki-44, Army Type 2 single-seat fighter Shoki (Demon). Fast but relatively unmanoeuvrable interceptor.
Tony	Kawasaki Ki-61, Army Type 3 fighter Hien (Swallow). Famous fighter type, initially thought to be a Messerschmitt Bf 109 derivative due to its inline engine, unusual on Japanese fighters.
Topsy	Mitsubishi L4M, Navy Type 0 transport. Ki-57, Army Type 100 transport. Twin-engined, eleven-seat military counterpart to pre-war MC-20 airliner, closely resembling **Sally** bomber.
Trixie	Name allocated to expected licence-built Junkers Ju 52/3m transport.
Trudy	Name allocated to expected licence-built Focke-Wulf Fw 200 Condor.
Val	Aichi D3A, Navy Type 99 carrier bomber. Classic Navy dive-bomber, used extensively at Pearl Harbor.
Willow	Yokosuka K5Y, Navy Type 93 intermediate trainer. Long-serving, radial-engined biplane.
Zeke	Mitsubishi A6M, Navy Type 0 carrier fighter. World-famous fighter, better known on both sides as the **Zero**.

159

APPENDIX 2
NATO/ASCC Reporting Names

After the war the USAF introduced a series of type numbers to identify Soviet aircraft, but the system soon fell into disrepute when it became apparent that some aircraft had been allocated more than one type number.

In 1954, the Air Standards Coordinating Committee (ASCC) devised a new series of NATO reporting names, beginning with B for Bombers, C for Cargo or Transport aircraft (including commercial airliners), F for Fighters, H for Helicopters, and M for Miscellaneous types including trainers; conversion trainer models of standard fighter types were initially given a separate name in the Miscellaneous category, but this practice was later discontinued. Around 1980, a new A for Attack category was projected to begin with what eventually emerged as the Su-25, but the plan was later discarded and the Su-25 became the Frogfoot in NATO parlance.

Titles are also sub-divided into single-syllable names for airscrew-driven aircraft (piston or turbine) and two-syllable names for jet-propelled aircraft; this distinction does not apply to helicopters, or of course to missiles, which have their own categories, as follows: A for Air-to-Air, G for Surface-to-Air (SAMs); K for Air-to-Surface, and S for Surface-to-Surface (ranging from ICBMs and SLBMs to anti-tank). Even radar, ECM systems, etc., were allocated reporting names, although the only apparent format was the use of two seemingly random words, leading to such curious titles as Puff Ball, Odd Rods, and even Big Nose.

NATO/ASCC names were also applied to Chinese aircraft, but as these were predominantly copies of Soviet types (for which the original ASCC name held) the only original names were Fantan and Finback.

The whole reason for the NATO/ASCC system (i.e. the obsessive Soviet secrecy during the Cold War) was perhaps best illustrated by the well-known incident during a US visit to Kubinka airfield in 1956, when a query about the designation of a particular bomber was met with the riposte: 'That is what you call the Bison'.

Cracks began appearing in the system in the seventies, when US intelligence sources began publishing names such as Ram J and Tag D for new types photographed by satellites at the Ramenskoye and Taganrog research establishments. With the emergence of glasnost in the late eighties, however, the whole *raison d'être* of the NATO/ASCC system gradually disappeared as more and more information about Soviet/CIS types became available to the West. The last name allocated appears to have been Madcap for the An-74, and it can only be speculated what names were left unused on the ASCC's list. Nowadays even Russians speak of Blackjacks and Flankers, and for the Western enthusiast a little of the wonder has perhaps been lost compared to the days of those surreptitiously obtained photos of intriguing new Soviet fighters with bizarre reporting names like 'Frogspawn'. A small price to pay, however.

Information in the tables includes NATO/ASCC reporting names, formal

APPENDIX 2

designations and names, early USAF type numbers (where applicable), Ram/Tag names (again where applicable) and brief notes including the year the aircraft was first flown or first seen. Design Bureau abbreviations are: An – Antonov, Be – Beriev (now the Taganrog Aviation Development & Manufacturing Enterprise), Che – Chetverikov, Il – Ilyushin, Ka – Kamov, La – Lavochkin, Li – Lisunov, MiG – Mikoyan & Gurevich, Mi – Mil, Mya – Myasischev, Pe – Petlyakov, Po – Polikarpov, Su – Sukhoi, Ts – Tsibin, Tu – Tupolev, Yak – Yakovlev.

Backfin	Tu-98. Experimental transonic medium bomber, twin engines in fuselage, fed by shoulder-mounted intakes. 1955.
Backfire	Tu-22M. Swing-wing supersonic bomber. Once thought in the West to be designated Tu-26 from confusion with prototype-only designation of Tu-126. 1971.
Badger	Tu-16. Type 39. Twin-jet medium bomber, engines in wing roots. 1951.
Bank	North American B-25 Mitchell supplied during WW2.
Barge	Tu-85. Type 31. Unsuccessful derivative of **Bull**. 1951
Bark	Il-2. Shturmovik ground-attack monoplane. 1939.
Bat	Tu-2. Twin-engined light bomber. 1941.
Beagle	Il-28. Type 27. Twin-jet light bomber. Counterpart to **Canberra** (cf **Butcher and Mascot**). 1948.
Bear	Tu-95 (Bear A to E) and Tu-142 (Bear F to J). Type 40. Once thought to have been Tu-20. Four-turboprop strategic bomber, swept wings and tail.
Beast	Il-10. Developed version of **Bark**. 1944.
Beauty	Tu-22. Original name for **Blinder**, reputedly stemming from misinterpreted remark by US official ('Gee, that's a beauty!') at 1961 Tushino flypast. Later discarded as too complimentary.
Bison	Mya-4 Molot (Hammer). Type 37. Heavy strategic bomber. 1954.
Blackjack	Tu-160. Ram P. Swing-wing supersonic strategic bomber. Similar in layout to Rockwell B-1B Lancer, but larger. 1981.
Blinder	Tu-22. Twin-jet supersonic bomber with engines at base of tail; formerly **Beauty**. **Blinder** name originally used for Tu-28 **Fiddler** when thought to be a bomber.
Blowlamp	Il-54. Swept-wing, twin-podded engines. Built to same specification as **Brewer**, but no production. 1954.
Bob	Il-4. Long-range, twin-engined medium bomber. 1935.
Boot	Tu-91. Single turboprop experimental ASW aircraft. 1956.
Bosun	Tu-14. Type 35. Torpedo/recce. Limited service. 1947.
Bounder	Mya-50. Poorly-designed, tailed-delta supersonic bomber with two engines in underwing pods and two on wingtips; did not enter service. 1959.
Box	Douglas A-20 Havoc supplied during WW2.
Brassard	Yak-28. Original name for **Brewer**, changed to avoid confusion with Max Holste Broussard utility aircraft.
Brawny	Il-40. Outmoded attempt to produce a jet Shturmovik type. Twin engines fed by nose intakes. 1953.

161

APPENDIX 2

Brewer	Yak-28. Swept-wing bomber, twin engines in underwing pods. 1960.
Buck	Pe-2. Three-seat, twin-engined light bomber. 1938.
Bull	Tu-4. Copy of Boeing B-29 Superfortress. 1947.
Butcher	Il-28/Tu-82. Initial name for Il-28 **Beagle**, also used for swept Tu-82 to same specification. 1948.
Cab	Li-2. Licence-built version of Douglas C-47; also applies to aircraft supplied by America during WW2.
Camber	Il-86. First Soviet widebody airliner. Max. 350 passengers. Four engines. 1976.
Camel	Tu-104. First Soviet jet airliner. Derived from Tu-16 **Badger**. 1955.
Camp	An-8. Twin-engined, high-wing 48-seater. Only 100 built. 1955.
Candid	Il-76. Heavy transport broadly similar in role and appearance to Lockheed C-141 StarLifter. 1971.
Careless	Tu-154. Three-engined, 150-seat, medium-to long-range airliner. 1968.
Cart	Tu-70. One-off 48-seat transport variant of Tu-4 **Bull**. 1946.
Cash	An-28. Stretched turboprop derivative of An-14 **Clod**, initially designated An-14M. Produced exclusively in Poland. 1969.
Cat	An-10 Ukraina (Ukraine). Four-engined development of An-8 **Camp**. Civil use only. 1957.
Charger	Tu-144. World's first supersonic transport, similar in appearance to Concorde. 1968.
Clam	Il-18. Long-range four-engined transport, one only. Not to be confused with later Il-18 **Coot**. 1947.
Clank	An-30. Photo-survey version of An-26 **Curl**. 1973.
Classic	Il-62. Long-haul four-jet airliner. VC-10 lookalike. 1962.
Cleat	Tu-114 Rossiya (Russia). Transport counterpart to Tu-95 **Bear**. Once the world's largest airliner. 1957.
Cline	An-32. Higher-powered derivative of An-26 **Curl** with engines mounted on upper wing surfaces. 1976.
Clod	An-14 Pchelka (Bee). Twin-engined utility transport. Protracted development. 1958.
Coach	Il-12. Low-wing, twin-engined transport. Over 2,000 built. 1946.
Coaler	An-72 (Coaler A & C), An-74 (Coaler B). STOL transport with twin turbofans mounted above wing leading-edge. An-74 is Arctic version. 1977.
Cock	An-22 Antei (Antheus). Four-turboprop heavy strategic freighter. 1965.
Codling	Yak-40. Tri-jet, straight-winged feederliner. 1966.
Coke	An-24. Twin-turboprop, high-wing transport. 1960.
Colt	An-2. Mass-produced utility biplane. 1947.
Condor	An-124 Ruslan. Heavy logistic freighter, comparable to Lockheed C-5 Galaxy. 1982.
Cooker	Tu-110. Stretched four-engined derivative of Tu-104 **Camel**. Prototype only. 1957.
Cookpot	Tu-124. Scaled-down, 56-seat version of Tu-104. 1960.
Coot	Il-18. Four-turboprop 120-seater (cf **May**). 1957.

162

Cork	Yak-16. Twin-engined ten- to twelve-seat feederliner. 1947.
Cossack	An-225 Mriya (Dream). Six-engined derivative of An-124 **Condor**. Currently world's largest aircraft. 1988.
Crate	Il-14. Re-designed version of Il-12 **Coach**. 1950.
Creek	Yak-12. High-wing cabin monoplane. 1946.
Crib	Yak-8. Eight-seat, low-wing feederliner. Few built. 1945.
Crow	Yak-10. Side-by-side trainer. One only. 1944.
Crusty	Tu-134. Short-haul, twin-jet airliner. 1962.
Cub	An-12. Tactical transport derivative of An-10 **Cat**. Analogous to Lockheed C-130 Hercules. 1959.
Cuff	Be-30. High-wing feederliner twin, abandoned in 1970.
Curl	An-26. Refined development of An-24 **Coke**. 1969.
Faceplate	MiG-21. Swept-wing version of famous MiG-21 **Fishbed**. Unsuccessful competitor to Su-7 **Fitter**. 1955.
Fagot	MiG-15. Type 14. Well-known swept-wing fighter, counterpart to North American F-86 Sabre. Built in China as Shenyang F-2. 1947.
Faithless	MiG-23DPD. Not to be confused with production MiG-23 **Flogger**. Experimental STOL fighter with fixed-wing, tailed-delta layout. 1967.
Falcon	MiG-15. Original name for MiG-15 **Fagot**.
Fang	La-11. Final Soviet piston-engined fighter. 1947.
Fantail	La-15. Type 21. High-wing jet fighter, used mainly in ground-attack role. 1948.
Fantan	Chinese Qiang-5 ground-attack derivative of MiG-19 **Farmer**, with 'solid' nose and side-mounted intakes. 1965.
Fargo	MiG-9. First Soviet jet fighter. 1946.
Farmer	MiG-19. Soviet Union's first true supersonic fighter. Built in China as Shenyang F-6. 1953.
Fearless	Name used for hypothetical swing-wing, twin-turbofan strike-fighter, used for comparative studies by Pentagon in 1971.
Feather	Yak-15 (Type 2) and Yak-17 (Type 16). Single-seat jet fighters based on piston Yak-3. 1946.
Fencer	Su-24. Swing-wing two-seat strike aircraft. Maritime recce version is Fencer E. 1970.
Ferret	'Mig-37'. Spurious title used for model kit of imaginary Soviet stealth fighter.
Fiddler	Tu-28. Heavy long-range interceptor (cf **Blinder**). 1957.
Fin	La-7. WW2 piston-engined fighter. 1943.
Finback	Indigenous Chinese F-8 jet fighter with tailed-delta layout and twin engines. Original F-8 (Finback A) had nose intake, but F-8-II (Finback B) has 'solid' radar nose and side-mounted intakes. 1969.
Firebar	Yak-28P. Interceptor variant of prolific Yak-28 series. 1960.
Fishbed	MiG-21. Long-running, multi-role fighter series. MiG-21bis is Fishbed N (cf **Mongol**). Also built in China as Chengdu F-7. 1955.
Fishpot	Su-9 (Fishpot A & B) and Su-11 (Fishpot C). Single-seat, tailed-delta interceptors with nose intake (cf **Maiden**). 1956.
Fitter	Su-7 (Fixed-wing Fitter A) and Su-17 -20, and 22 (swing-wing Fitter C to K). Single-seat, ground-attack aircraft (cf **Moujik**). 1956.

Flagon	Su-15. Tailed-delta interceptor. Twin engines and long radar nose. 1965.
Flanker	Su-27. Ram K. Twin-tailed, counter air fighter. Variants are two-seat Su-27IB (Su-34) strike aircraft, Su-27K or Su-33 naval version with foreplanes, and Su-35 advanced air superiority fighter. 1977.
Flashlight	Yak-25. Two-seat, all-weather interceptor. Twin engines in underwing pods. 1952.
Flipper	MiG Ye-152A. Experimental fighter resembling a scaled-up, twin-engined MiG-21 **Fishbed**. 1961.
Flogger	MiG-23 (Flogger A, B, C, E, F, G, H, K) and MiG-27 (Flogger D & J). Single-engined, swing-wing fighters; Mig-27 is specialised ground-attack variant.
Flora	Yak-23. Type 28. Clear-weather interceptor, similar in layout to Yak-15/17 **Feather**. Used in small numbers. 1947.
Forger	Yak-38. Shipborne VTOL fighter, originally designated Yak-36MP. One vectored-thrust engine and two lift engines. 1971.
Foxbat	Mig-25. Twin-tailed Mach 2.8 interceptor. 1964.
Foxhound	MiG-31. Two-seat derivative of MiG-25 **Foxbat** with more advanced weapons system. Initially MiG-25M. 1975.
Frank	Yak-9. WW2-vintage piston fighter.
Fred	Bell P-39 Airacobra supplied during WW2.
Freehand	Yak-36. Single-seat, vectored-thrust VTOL research aircraft. 1967.
Freestyle	Yak-141. Ram T. Supersonic V/STOL shipboard fighter. One vectored-thrust engine and two lift engines. Prototypes only. 1989.
Fresco	MiG-17. Type 20. Development of MiG-15 **Fagot**, also widely used. 1949.
Fritz	La-9. All-metal successor to La-5 and La-7 series. 1946.
Frogfoot	Su-25. Ram J. Twin-engined close support aircraft. Su-28 is export variant. 1977.
Frosty	Title applied to supposed Tu-10 fighter in the mid-fifties, now believed not to have existed.
Fulcrum	MiG-29. Ram L. Single-seat, twin-engined, counter air fighter; MiG-33 is improved version. 1977.
Halo	Mi-26. Heavy-lift helicopter with cargo-hold similar in size to Lockheed C-130 Hercules. 1977.
Hare	Mi-1. Widely used four-seat utility helicopter. 1948.
Harke	Mi-10. Flying crane derivative of Mi-6 **Hook**. 1960.
Harp	Ka-20. Pre-production model of Ka-25 **Hormone**. 1961.
Hat	Ka-10M. So-called flying motorcycle. Four only built. 1954.
Havoc	Mi-28. Two-seat anti-armour/attack helicopter. 1982.
Haze	Mi-14. Amphibious derivative of Mi-8 **Hip**. 1973.
Helix	Ka-27, Ka-29 (Helix A, B, D) and Ka-32 (Helix C). Multi-role, twin-turbine helicopter. Successor to Ka-25 **Hormone** series. Ka-28 is export variant. 1974.
Hen	Ka-15. Two-seat, light utility/ASW helicopter. 1952.
Hermit	Mi-34. Light, piston-engined training helicopter. 1986.
Hind	Mi-24. Assault and anti-armour helicopter. Mi-25 and Mi-35 are export variants. 1972.

Hip	Mi-8 and Mi-17 (Hip H). Widely used twin-turbine transport helicopter. Mi-17 is improved version. 1961.
Hog	Ka-18. Four-seater derived from Ka-15 **Hen**. 1957.
Hokum	Ka-50 Werewolf. Single-seat attack helicopter. 1982.
Homer	Mi-12. Heavy, twin-rotor, outrigged helicopter. 1968.
Hoodlum	Ka-26. Light utility helicopter. Turbine version built in Romania is Ka-126 (Hoodlum B). 1965.
Hook	Mi-6. Once world's largest helicopter, seating up to 90 in passenger form. 1957.
Hoop	Ka-22 Vintokryl (screw-wing) compound helicopter to same broad principle as Fairey Rotodyne, but with twin outrigged rotors. Experimental only. 1961.
Hoplite	Mi-2. Twin-turbine development of Mi-1 **Hare**. 1961.
Hormone	Ka-25. Naval ASW helicopter. 1961.
Horse	Yak-24. Type 38. Tandem-rotor transport used in limited numbers. 1952.
Hound	Mi-4. Type 36. General-purpose helicopter, roughly similar in appearance to Sikorsky S-55. 1953.
Madcap	An-74. Airborne Early Warning & Control variant of An-74 **Coaler** with rotodome on top of fin. Project cancelled 1990.
Madge	Be-6. Twin-engined ASW flying boat. 1949.
Maestro	Yak-28U. Conversion trainer model of Yak-28 series. 1962.
Magnet	Yak-17UTI. Type 26. Training variant of **Feather**. 1948.
Magnum	Yak-30. Tandem basic trainer, lost out to **Maya**. 1960.
Maiden	Su-9UTI. Conversion trainer model of Su-9 **Fishpot**. 1957.
Mail	Be-12 Chaika (Seagull). Twin-turboprop ASW flying boat. 1961.
Mainstay	Beriev A-50. Airborne Early Warning & Control variant of Ilyushin Il-76 **Candid**. 1979.
Mallow	Be-10. High-wing jet flying boat, few built. 1960.
Mandrake	Yak-25RV. High-altitude recce variant of Yak-25 with new straight wings. 1953.
Mangrove	Yak-27R. Tactical recce model of Yak-27 series. 1958.
Mantis	Yak-32. Single-seat version of Yak-30 **Magnum**. Two only. 1960.
Mare	Yak-14. Type 24. Heavy assault glider with folding nose. 1949.
Mark	Yak-7U. Trainer version of WW2 Yak-7 fighter. 1940.
Mascot	Il-28U. Type 30. Trainer version of Il-28 **Beagle**. 1949.
Max	Yak-18. Classic radial-engined trainer, in production over 25 years. 1946.
May	Il-38. Maritime patrol/ASW variant of Il-18 **Coot** transport, 1967.
Maya	Aero (Czech) L-29 Delfin (Dolphin). Standard Warsaw Pact basic trainer for many years. 1959.
Mermaid	Be-40 (or A-40) Tag D. Twin-turbofan amphibious ASW aircraft. Be-42 is air-sea rescue version. 1987.
Midas	Il-78. Tanker variant of Il-76 **Candid**. 1985.
Midget	MiG-15UTI. Type 29. Conversion and advanced training version of MiG-15 **Fagot**. 1949.
Mink	Yak UT-2. Low-wing, fixed-undercarriage trainer. 1934.
Mist	Ts-25. Type 25. Experimental transport glider. 1945.

Mole	Be-8. Type 33. Single-engined utility amphibian. 1947.
Mongol	MiG-21UTI. Conversion trainer models of **Fishbed**. 1958.
Moose	Yak-11. Basic trainer, replaced by **Maya**. 1947.
Mop	Consolidated Catalina supplied during WW2. Also applies to licence-built GST variant.
Moss	Tu-126. AWACS derivative of Tu-114 **Cleat**. 1969.
Mote	Beriev MBR2. Light, pusher-engined flying boat. 1934.
Moujik	Su-7UTI. Conversion trainer model of fixed-wing **Fitter** series. 1957.
Mouse	Supposedly a modified version of Yak-18 designated Yak-18M, but may never have existed.
Mug	Che-2. Twin-engined flying boat. 1937.
Mule	Po-2. Mass-produced trainer/multi-purpose biplane. 1928.
Mystic	Mya M-17. Ram M. Twin-boom, high-altitude research aircraft. Original Stratospherica version was single-engined, M-55 Geophysica version is twin-engined. 1982.

MISSILES

Air-to-Air

Acrid	AA-6
Alamo	AA-10
Alkali	AA-1
Amos	AA-9
Anab	AA-3
Apex	AA-7
Aphid	AA-8
Archer	AA-11
Ash	AA-5
Atoll	AA-2

Surface-to-Air

Gainful	SA-6
Gammon	SA-5 (originally Griffon)
Ganef	SA-4
Gaskin	SA-9
Gecko	SA-8
Gladiator	SA-12
Goa	SA-3 & naval SA-N-1
Gopher	SA-13
Grail	SA-7 & naval SA-N-7
Gremlin	SA-14
Grumble	SA-10
Guideline	SA-2
Guild	SA-1

Air-to-Surface

Kangaraoo	AS-3
Karen	AS-10
Kedge	AS-14
Kegler	AS-12
Kelt	AS-5
Kennel	AS-1
Kent	AS-15
Kerry	AS-7
Kickback	AS-16
Kilter	AS-11
Kingbolt	AS-13
Kipper	AS-2
Kitchen	AS-4
Krypton	AS-17

Surface-to-Surface
(AT = Anti-Tank; SS-N = Naval)

Saddler	SS-7
Sagger	AT-3 (Miliutka)
Sandal	SS-4
Sandbox	SS-N-12
Sapwood	SS-6
Sark	SS-N-4
Sasin	SS-8
Savage	SS-13

Sawfly	SS-N-6	**Shyster**	SS-3
Scalpel	SS-24	**Sibling**	SS-2
Scaleboard	SS-12	**Sickle**	SS-25
Scapegoat	SS-14	**Siren**	SS-N-9
Scarab	SS-21	**Skean**	SS-5
Scarp	SS-9	**Snapper**	AT-1 (Shmell)
Scrag	SS-10	**Spandrel**	AT-5
Scrooge	SS-15	**Spider**	SS-23
Scud	SS-1	**Spigot**	AT-4
Scrubber	SS-N-1	**Spiral**	AT-6
Sego	SS-11	**Stiletto**	SS-19
Serb	SS-N-5	**Styx**	SS-N-2
Shaddock	SS-N-3	**Sunburn**	SS-N-22
Shipwreck	SS-N-19	**Swatter**	AT-2

APPENDIX 3

Formal British Military Naming Systems

February 1918 System

The first known suggestion of a formal procedure for naming British military aircraft came in 1916, with a simple proposal by the Parliamentary Air Committee to name RFC types after land birds and RNAS types after sea birds. The idea was not taken up, but two years later the Department of Aircraft Production introduced a basically alliterative system whereby each company's products would have a name beginning with a set of specified letters as follows (with examples):

Airco	Am, Ab (Amiens)
Armstrong Whitworth	Ar, Aw (Armadillo)
Austin	As, Au, Os (Osprey)
Blackburn	Bl, Bu (Blackburd)
Boulton & Paul	Bo (Bobolink)
British Aerial Transport	Ba (Baboon)
Bristol	Br, Co, Ko :Braemar)
Fairey	Fa, Pha (Fawn)
Grahame-White	Ga (Ganymede)
Grain/Port Victoria	Gr (Griffin)
Handley Page	H (Hanley)
Nieuport	Ni (Nighthawk)
Parnall	Pa (Panther)
Royal Aircraft Factory	Ra (Ram)
Shorts	Sh (Shamrock)
Siddeley	Si, Cy, Si (Siskin)
Sopwith	Sa, Sn, So (Snail)
Supermarine	Su, Sw (Swan)
Vickers	U, V (Victoria)
Westland	Wa, We, Ye (Walrus)

The scope of the system was severely restricted, however, by the further requirements that each name should also comply with both of the following name and weight categories:

Single-engine, single-seat	Land birds, Insects and Reptiles
Single-engine, multi-seat	Mammals
Two-engines	Towns and Cities
Three engines	Historical or Mythological names (masculine)

168

APPENDIX 3

Four engines

Single-engine seaplanes
Twin-engine seaplanes
Three-engine seaplanes
Four-engine seaplanes
0 to 5 tons weight
5 to 10 tons weight
10 to 20 tons weight
20 tons weight or above

Geographical features and place
names
Fish and Waterfowl
Coastal Towns
Rivers and Female names
Islands and Lakes
French names
Other European names
British names
African and Asian names

July 1918 System

Predictably, the February 1918 System quickly proved too restrictive to sustain,
and was amended just five months later. Much of the earlier method remained
intact, but the alliterative requirement was dropped (several companies continued
it anyway), foreign names were excluded, and names of cats, birds of prey and
snakes were reserved for aero-engines. The full range of categories (with
examples where available) was:

Single-engine, single-seat
land planes

Single-engine, multi-seat
Multi-engine landplanes up to
11,000lb weight
Multi-engine landplanes between
11,000lb and 20,000lb weight
Multi-engine landplanes between
20,000lb and 45,000lb weight
Multi-engine landplanes above
45,000lb weight
Single-engine, multi-seat seaplanes
Multi-engine seaplanes up to 11,000lb
weight
Multi-engine seaplanes between
11,000lb and 20,000lb weight
Multi-engine seaplanes between
20,000lb and 45,000lb weight
Multi-engine seaplanes above
45,000lb weight

Reptiles, Land Birds and Insects
(Westland Tadpole, Sopwith Cuckoo,
Avro Spider)
Mammals (Avro Bison)
Inland towns in England and Wales
(Nieuport London)
Inland towns in Scotland and Ireland
(Bristol Braemar)
Mythological names (no known
examples)
'Attribute' names ending in '-ant',
'-ent' or '-ous' (no known examples)
Fish and Waterfowl (Fairey Pintail)
Coastal towns in England and Wales
(Short Cromarty)
Coastal towns in Ireland and Scotland
(Phoenix Cork)
Historical or Mythological names (no
known examples)
'Attribute' names ending in '-al', '-er',
or '-ic' (English Electric Eclectic)

1921 System

The 1921 System, administered by the Directorate of Research, was again based
largely on weight, and for the first and only time sought (none too successfully)

169

to include commercial airliners:

Single-seat landplanes	Land birds (Gloster Gamecock)
Multi-seat landplanes up to 6,000lb weight	Mammals (Hawker Hart)
Multi-seat landplanes between 6,000lb and 12,000lb weight	Inland British Towns (Handley Page Hampstead)
Multi-seat landplanes between 12,000lb and 20,000lb weight	Inland British Empire Towns (Handley Page Hyderabad)
Multi-seat landplanes between 20,000lb and 40,000lb weight	Mythological names (no known examples)
Multi-seat landplanes over 40,000lb weight	'Attribute' names ending in '-ant', '-ent' or '-ous' (no known examples)
Single-seat seaplanes	Shellfish (Short Cockle)
Multi-seat seaplanes up to 6,000lb weight	Fish (Blackburn Sprat)
Multi-seat seaplanes between 6,000lb and 12,000lb weight	British Coastal Towns (English Electric Ayr)
Multi-seat seaplanes between 12,000lb and 20,000lb	Coastal British Empire Towns (Blackburn Sydney)
Multi-seat seaplanes between 20,000lb and 40,000lb	Mythological names (Saro Valkyrie)
Multi-seat seaplanes above 40,000lb weight	'Attribute' names ending in '-al', '-er' or '-ic' (no known examples)
Civil aircraft	Historical names, excluding personalities (Handley Page Hamlet)
Civil seaplanes	Historical personalities (no known examples)

1927 System

This System was the simplest of all, specifying only the initial letter for each category of aircraft:

Army Co-op (A)	(Hawker Audax)
Bombers, single-engine (P)	(no known examples)
Bombers, multi-engined (B)	(no known examples)
Bombers, torpedo (M)	(no known examples)
Coastal Reconnaissance (R)	(Short Rangoon)
Fighters, land (F)	(Hawker Fury)
Fighters, naval (N)	(Hawker Nimrod)
Fighters, reconnaissance (O)	(Hawker Osprey)
General-Purpose (G)	(Fairey Gordon)
Spotters, naval (S)	(Fairey Seal)
Trainers (T)	(Hawker Tomtit)
Troop Carriers (C)	(Handley Page Clive)

1932 System

The naming system was enlarged again in 1932, and led to some of the great names in British aviation:

Army Co-op	Classical names (Westland Lysander)
Bombers, day	Animals (Hawker Hind)
Bombers, night	Inland British Empire Towns, or towns with RAF connections (Vickers Wellington, Fairey Hendon)
Fighters, land	Names suggesting activity, aggression, speed, etc. (Supermarine Spitfire)
Fighters, naval	Mythological names (Blackburn Roc)
Fighter Recce, Naval	Sea Birds (Fairey Fulmar)
Flying Boats	Coastal British Empire Towns (Blackburn Perth)
General-Purpose	Historic British names (Hawker Hardy)
Spotters, naval	Marine Animals (Supermarine Walrus)
Torpedo Bombers	Seas and Waterways (Blackburn Baffin)
Trainers	Names connected with education (Airspeed Oxford)
Transports	British Empire Towns (Vickers Valentia)

1939 System

The 1939 system was broadly similar to its predecessor but with the Fleet Air Arm becoming independent from the RAF, responsibility for RAF and RN aircraft names passed to the Ministry of Aircraft Production and the Admiralty respectively. The latter seems to have used no strict formula, but instead used 'Sea' prefixes for navalised landplanes (e.g. Sea Gladiator, Sea Hurricane) or ad hoc names of a general marine nature (e.g. Fairey Barracuda, Supermarine Sea Otter, etc). American types introduced to either service were usually renamed, but with the standardisation of titles in 1944 most reverted to their original American names.

Army Co-op	Classical names (no examples known)
Bombers (day/night)	Inland British Empire Towns (Avro Lancaster)
Fighters	Names suggesting activity, aggression, speed, etc. (Westland Whirlwind)
Flying Boats	Coastal towns (Short Shetland)
General-Purpose, inc. torpedo bomber landplanes	Historical British names (Armstrong Whitworth Albemarle)

APPENDIX 3

Gliders	Names of historical military leaders beginning with H (Waco Hadrian)
Trainers	Names connected with education (Airspeed Cambridge)
Transports	British counties (de Havilland Hertfordshire): later expanded to include towns (Avro York)

Among names adopted for American types were the following (details can be found in the main section of the book): Argus, Bermuda, Boston, Caribou, Chesapeake, Dakota, Gannet, Gosling, Hadrian, Havoc, Hoverfly, Kittyhawk, Martlet, Maryland, Mohawk, Nomad, Tarpon.

Since the war the number of types in service at any one time has steadily diminished, and with it has gone the need for any formal naming procedure. The traditions lingered on with such as the Percival Provost, de Havilland Devon, and English Electric Lightning, but gradually more and more types already came with names from abroad (e.g. Sabre, Phantom, Hercules) or used the names of their commercial forebears (e.g. Comet and Britannia). By the mid-sixties the naming of RAF aircraft had become so patchy that of three commercially-derived transports entering service within two years of each other, the HS 748MF was given the traditional-style name of Andover, the Beagle B206 became the Basset apparently to comply with the maker's 'Kennel Club' series, and no consideration whatsoever seems to have been given to naming the Vickers/BAC VC10.

Now there seems to be no pattern of names for British military aircraft, with each case being considered on its own merits. With international collaboration very much the order of the day, consideration must of course be given to names which will not offend former adversaries, and which are acceptable in more than one language.

APPENDIX 4

Series Names

Many aircraft manufacturers use informal themes for naming their aircraft. The following lists should not be regarded as exhaustive, but in many cases the names can also be found in the main section of the book.

Airspeed (UK)
Diplomatic Corps names: Ambassador, Consul, Courier, Envoy, Viceroy.

Beagle (UK)
'Kennel Club' names: Airedale, Basset, Bulldog, Pup, Terrier.

Beech (US)
Aristocracy/Royalty: Baron, Duchess, Duke, King Air, Queen Air.

Bell (US)
Initially, names beginning with 'Aira': Airabonita, Airacobra, Airacomet, Airacuda.

Boeing (US)
'Strato' names: Stratocruiser, Stratofortress, Stratofreighter, Stratojet, Stratoliner, Stratotanker.

Curtiss (US)
Bird names, notably Condor for transports and Hawk (and variations thereof) for fighters. Shrike and Helldiver were trade names for Army/Navy attack aircraft, but were not always formally recognised (see main section of book).

De Havilland Canada
Canadian animals: Beaver, Buffalo, Caribou, Chipmunk, Otter.

Douglas (US)
Names beginning with Sky- or ending in -Master: Skyhawk, Skyknight, Skylancer, Skymaster, Skypirate, Skyraider, Skyray, Skyrocket, Skystreak, Skywarrior; Cargomaster, Globemaster, Mixmaster.

Focke-Wulf (Germany)
Bird names: Condor, Falke (Falcon), Mowe (Seagull), Owl, Stieglitz (Goldfinch), Weihe (Kite).

Grumman (US)
'Fighting Felines' and seabirds: Bearcat, Cougar, Hellcat, Jaguar, Panther, Tiger, Tigercat, Tomcat, Wildcat; also Kitten and Ag-Cat; Albatross, Duck, Goose, Mallard, Pelican, Widgeon.

APPENDIX 4

Lockheed (US)

McDonnell (US)

Northrop (US)

Piper (US)

Republic (US)

Sopwith (UK)

Astronomy/Heavenly Bodies:
Constellation, Galaxy, Lodestar,
Orion, Saturn, Starfighter, Starlifter.
Supernatural phenomena: Banshee,
Demon, Goblin, Phantom, Voodoo.
Early types, letters of Greek alphabet:
Alpha, Beta, Delta, Gamma.
Indian names: Apache, Aztec,
Cherokee, Cheyenne, Comanche,
Mojave, Navajo, Pawnee, Seminole,
Seneca, Sequoya.
'Thundercraft': Thunderbolt,
Thunderchief, Thunderflash,
Thunderjet, Thunderstreak,
Thunderwarrior.
'Menagerie' names: Antelope,
Buffalo, Camel, Dolphin, Gnu,
Hippo, Salamander, Snail, Wallaby,
Rhino.

174

APPENDIX 5

Phonetic Alphabets

As well as their primary function of helping to avoid ambiguity in radio transmissions, phonetic alphabet words appear in a number of other aviation contexts. RAF wartime bombers were known within their squadrons as 'G-George' etc. US fighter and cargo transport aircraft are occasionally referred to as (for example) 'Fox-4s' or 'Charlie-130s' (Phantoms and Hercules) and aircraft designations can even lead to a phonetically-based nickname, such as Able Dog for the Douglas AD Skyraider and H-Pip for WW1 Handley Page bombers. Also from the old WW1 alphabet comes 'Ack-Ack' for anti-aircraft, but the other classic example from that era is actually a corruption of the system: air mechanics in the RFC were widely known as Ack Emmas (the sound 'Emm' being extended to Emma) possibly because Monkey was considered offensive. Another common use of phonetic alphabets is to enunciate suffix letters for different marks of an aircraft, as in the Bf 109E Emil, Ju 87R Richard, Hotel-model Huey (UH-1H) and Saab J-35B Bertil. The current ICAO alphabet came into effect on 1 March 1956.

	British, pre-WW1	Allied from 1943	German, WW2	Current
A	Ack	Able	Anton	Alpha
B	Beer	Baker	Berta	Bravo
C	Charlie	Charlie	Casar	Charlie
D	Don	Dog	Dora	Delta
E	Edward	Easy	Emil	Echo
F	Freddie	Fox	Fritz	Foxtrot
G	George	George	Gustav	Golf
H	Harry	How	Heinrich	Hotel
I	Ink	Item	Ida	India
J	Johnnie	Jig	Josef	Juliet
K	King	King	Karl	Kilo
L	London	Lore	Ludwig	Lima
M	Monkey	Mike	Martha	Mike
N	Nuts	Nun	Nordpol	November
O	Orange	Oboe	Otto	Oscar
P	Pip	Peter	Paula	Papa
Q	Queen	Queen	Quelle	Quebec
R	Roger	Roger	Richard	Romeo
S	Sugar	Sugar	Siegfried	Sierra
T	Toc	Tare	Toni	Tango
U	Uncle	Uncle	Ulrich	Uniform
V	Vic	Victor	Viktor	Victor

W	William	William	Wilhelm	Whiskey
X	X-Ray	X-Ray	Xantippe	X-Ray
Y	Yorker	Yoke	Ypern	Yankee
Z	Zebra	Zebra	Zeppelin	Zulu

Czechoslovakia

A	Adam	B	Božewa	C	Cyril	D	David
Ď	Ďumbier	E	Emil	F	František	G	Gustav
H	Helena	CH	Chrudim	I	Ivan	J	Josef
K	Karol	L	Ludvik	Ĺ	Ĺubochňa	M	Maria
N	Norbert	Ň	Ňitra	O	Oto	P	Peter
Q	Quido	R	Rudolf	Ř	Řehor	S	Svatopluk
Š	Šimon	T	Tomas	Ť	Ťeplá	U	Urban
V	Václav	X	Xaver	Z	Zuzana	Ž	Žofia

(The Czech alphabet has no direct equivalent to the letters W or Y)

Poland

A	Adam	B	Barbara	C	Celina	D	Dorota
E	Ewa	F	Franciszek	G	Genowefa	H	Henryk
I	Irena	J	Jadwiga	K	Karol	L	Leon
M	Maria	N	Natalia	O	Olga	P	Pawel
R	Roman	S	Stanislaw	T	Tadeusz	U	Urszula
W	Waclaw	X	Xantypa	Z	Zygmunt		

(The Polish alphabet has no direct equvalents to the letters Q, V or Y)

Soviet Union/Russia

A	Az	B	Buki	V	Vedi
G	Glagol'	D	Dobro	Ye	Yest'
Zh	Zhivete	Z	Zemlya	I	Izhe
K	Ka	L	Lyudi	M	M'islete
N	Nash	O	On	P	Pokoi
R	Rts'i	S	Slovo	T	Tverdo
U	Ukho	F	Fert	Kh	Kha
Ts	Tsepochka	Ch	Cherv'	Sh	Shapka
Shch	Shcha	i	Yer'i	E	E'oborotnoye
Yu	Yula	Ya	Yako		

APPENDIX 6

Nicknames of US Navy Aircraft Carriers

Battlestar	USS Carl Vinson (CVN-70).
	From *Battlestar Galactica* TV series.
Big E	USS Enterprise (CV-6 & CVN-65)
Big John	USS John F. Kennedy (CVA-67)
Big O	USS Oriskany (CVA-34)
Bonny Dick	USS Bon Homme Richard (CVA-31)
Champ	USS Lake Champlain (CVA-39)
Connie	USS Constellation (CVA-64)
FDR	USS Franklin D. Roosevelt (CVA-42)
Fighting I	USS Intrepid (CVA-11)
Hannah/Fighting Hannah	USS Hancock (CVA-19)
Happy Valley	USS Valley Forge (CVA-45)
Ike	USS Dwight D. Eisenhower (CVN-69)
Indy	USS Independence (CVA-62)
Kitty	USS Kittyhawk (CVA-63)
Lady Lex	USS Lexington (CV-2 & CV-16)
Phil Sea	USS Philippine Sea (CVA-47)
Sara	USS Saratoga (CV-3)
Starship Enterprise	USS Enterprise (CVN-65).
	From *Star Trek* TV series.
Tico	USS Ticonderoga (CV-14)

GLOSSARY

AEW	Airborne Early Warning
AFB	Air Force Base
ATA	Air Transport Auxiliary
BAC	British Aircraft Corporation
BAe	British Aerospace
BEA	British European Airways
BOAC	British Overseas Airways Corporation
BSAAC	British South American Airways Corporation
CMA	Cie des Messageries Aeriennes
COIN	Counter-Insurgency
CTP	Chief Test Pilot
Elint	Electronic Intelligence
ENAER	Empresa Nacional de Aeronautica de Chile
ESM	Electronic Surveillance/Support Measures
GE	General Electric
HS	Hawker Siddeley
IAF	Indian Air Force
MAW	Military Airlift Wing
NACA	National Advisory Committee for Aeronautics
NASA	National Aeronautics & Space Administration
NATO	North Atlantic Treaty Organisation
P&W	Pratt & Whitney
P&WC	Pratt & Whitney Canada
RAAF	Royal Australian Air Force
RAF	Royal Air Force
RCAF	Royal Canadian Air Force
RFC	Royal Flying Corps
RLM	Reichsluftfahrtministerium (German State Ministry of Aviation)
RNAS	Royal Naval Air Service
RNZAF	Royal New Zealand Air Force
RR	Rolls-Royce
SAAF	South African Air Force
SAC	Strategic Air Command
SAM	Surface-to-Air Missile
SATIC	Super Airbus Transporter International Co.
SEA	South-East Asia
SFECMAS	Societe Francaise d'Etudes et de Constructions de Materiels Aeronautiques Speciaux
SFERMA	Societe Francais d'Entretien et de Reparation de Materiel Aeronautique

GLOSSARY

Sigint	Signals Intelligence
SLAR	Side-Looking Airborne Radar
SNCAC	Societe Nationale de Constructions Aeronautiques du Centre
SNCAN	Societe Nationale de Constructions Aeronautiques du Nord
SNCASE	Societe Nationale de Constructions Aeronautiques du Sud-Est
SNCASO	Societe Nationale de Constructions Aeronautiques du Sud-Ouest
SOE	Special Operations Executive
STOL	Short Take-Off & Landing
TARPS	Tactical Airborne Reconnaissance Pod System
TOW	Tube-launched Optically-tracked Wire-guided
USAAC	United States Army Air Corps
USAAF	United States Army Air Force
USAF	United States Air Force
USCG	United States Coast Guard
USN	United States Navy
VTOL	Vertical Take-Off & Landing

INDEX

The entries in this index are names referred to in the text which do not form separate entries.

Able Dog 121
Ace 60
Ack-W 17
Admiral class 136
Admiralty Air Dept.
  A.D. Scout 121
Adnan 56
A.E.G. R.I 56
Aerial Coupé 52
Aerial 2 82
Aeritalia (see also Fiat
  and Alenia) G 91
  56
Aermacchi (see also
  Macchi)
  MB 326 70
  MB 339 141
Aero L-29 Delfin 165
Aerobile 6
Aeroguayin 20
Aeronca
  C-2, C-3 Collegian
    14
  Model 65TC
  Defender 61
Aeroplane
  magazine 26, 80,
    130
Aeroscout 78
Aerospatiale
  AS 350 Ecuriel 6
  AS 365 Dauphin 42
  Fouga 90 147
  SA 316 Alouette III
    31
  SA 330 Puma 135
  SA 341 Gazelle 97
  SE 313 Alouette II
    80
  TB-30 Epsilon 95
Aggie 48
Aggressor 72
Agusta 57, 68, 113
  A 109 66
  SF-260 144
Ahit 12
Aichi
  AI-104 156
  B7A Ryusei 155
  D1A 159
  D3A 159
  E10A 155
  E11A 156
  E13A 156
  E16A Zuiun 158
Airabonita 83
Airbus 147
Airbus Industrie
  A300 106
  A310 45
  A320 119
Air Camel 64
Airco (see de Havilland)
Air Ferry 147
Airfox 31
Air France 36, 104
Airguard 10
Airknocker 14

Airlifter 65, 86, 146
Airliner 106, 141
Airspeed 173
  AS 10, AS 46
    Oxford 96
  AS 30 Queen Wasp
    33
  AS 45 Cambridge
    128
  AS 57 Ambassador
    83
Ajeet 111
Akikusa 52
Aladin 137
Albatross 111
Albatros D III 141
Alcock, John 92
Alenia G 222 122
Al Kahira 100
Allegheny Airlines 46,
  88, 91
Alliance P 1 107
Alligator 4
AllStar 7
Alo 31
Alpha 31
Aluminium Overcast
  25, 98
Ambrosini Supersette
  54
American Airlines 9,
  43, 51
American Jet Industries
  86, 136, 139
Amiot
  AAC 1 Toucan 8
  Type 122 49
Anahuac 6
Analogue 9
Anasal 4
Anatra D, DS 4
Andover 24
Angel 43, 66
Angel of Okinawa 16
Annie 48
Antonov
  An-2 82, 162
  An-8 145, 162
  An-10 Ukraina 162
  An-12 163
  An-14 Pchelka 162
  An-22 Antei 162
  An-24 162
  An-26 163
  An-28 112, 162
  An-30 119, 162
  An-32 129, 162
  An-72, An-74 162,
    165
  An-124 Ruslan 17,
    162
  An-225 Mriya 163
Anvil, Flying 120
Arado
  Ar 198 5
  Ar 232 90
  Ar 234 Blitz 93
Arcturus 96

Argentina class 103
Argonaut 94
Argus 146
Aries 3, 96
Ariete 54
Armstrong Whitworth
  Atlas 3
  AW 38 Whitley 13
  AW 55 Apollo 2
  AW650 Argosy 147
  FK3, FK8 17
  Tadpole 93
Arpin A-1 112
Arrow 5
ARV Aviation Super 2
  65
Ashton 129
Astra 35
Astrojet/AstroLiner
  51
Atlanta 53
Atlantic 4
Atlas Aircraft
  CHS-2 Rooivalk 135
  C4M Kudu 20
  Impala 70
  XH-1 Alpha 31
ATR 42, ATR 72 34
Aucan 131
Aurora 96
Auster 70
Austrian Aviatik
  B III 111
  CI, D I 17
Avenger 87
Avia
  B 534 41
  S-199, CS-199 7
Avocado 122
Avon 2
Avro
  Autogiros 111
  Roe 1 27
  Type 504 6
  Type 621 Tutor 36
  Type 652A Anson 47
  Type 683 Lancaster
    80
  Type 688, 689 Tudor
    129
  Type 694 Lincoln 82
  Type 696 Shackleton
    115
  Type 698 Vulcan 3,
    141
  Type 748 24
Avro Canada; CF-100
  Canuck 33
Ayres Thrush 143
Ayrshire 83

Baby Jet 48, 139
Baby Jumbo 74
BAC (see also British
  Aerospace)
  Canberra 35
  Concorde 35
  Lightning 55

One-Eleven 9
  Strikemaster 21
  TSR-2 137
Backstagger, Flying
  124
BAe (see British
  Aerospace)
Baghdad 55
Bahadur 38
Balsa Bomber 92
Baltimore Whore 148
Bamboo Bomber 110
Bamel 85
Banana, Flying 95
Bandeirulha 12
Banshee 12
Barak 76
Baronet 17
Barougan 137
Basketweave Bomber
  149
Bat 91
Batship 68
Baz 140
Bazant 76
B-Dash-Crash 148
BEA 8, 10, 33, 42, 51,
  59, 62, 77, 136, 142
Beagle 173
  B 121 Pup 26
  B 206 14
  M 218 86
Beagle 140
Bearcat 71
Beardmore Inflexible
  71
Bear Hunter 115
Beast 33, 91, 150
Beau 145
Beautiful Death 26
Beebee 134
Beech 173
  AT-10 Wichita 27
  Model 17 124
  Model 18 46
  Model 23 93
  Model 35 Bonanza
    39
  Model 50 Twin
    Bonanza 131
  Model 55 Baron 9
  Model 65 106
  Model 90 King Air
    140
  Model 95 Travel Air
    9
  T-34 Mentor 79
  XA-38 62
Beer Barrel 13
Belalang 53
Bellairus 57
Bell 173
  AH-1 Cobra 120
  F-109 30
  Model 47 57, 77
  Model 206 JetRanger,
    LongRanger 77
  Model 222 3

P-39 Airacobra 83,
  164
P-63 Kingcobra 101
UH-1 Iroquois 3, 67
XP-59 Airacomet 69
XP-77 138
Belvedere 25
Bereznyak-Isaev BI 40
Beriev
  A-40 165
  A-50 56, 165
  Be-6 165
  Be-8 166
  Be-10 165
  Be-12 Chaika 165
  Be-30 3, 163
  Be-40 165
  MBR-2 166
Berline 81
Bermuda 20
Bermuda class 103
Bertha 7
Bertil 2
Beta 135
Bev 124
Biafran Baby 89
Biff 23
Big Baltic 60
Big Charlie 113
Big Dipper 135
Big Gun 121
BigLifter 68
Big Mother 113
Big Stick 98
Big Top 74
Big White Bird 142
Bird 79
Biscayne 26 132
Biscuit Bomber 58
Bisley 20
BiStar 135
Blackbird 65
Blackburn
  B-1 Segrave 87
  B-24 Skua 112
  B-26 Botha 23
  B-48 50
  B-101 Beverley 124
  B-103 Buccaneer 11
  CA 15C 44
  R 1 Blackburn 26
  RB 1 Iris 106
Black Crow 138
Black Death 75, 117
Blackfish 127
Black Jet 150
Black Knight 127
Blériot 99
  Monoplanes 18
Blind Bat 65
Blind Wonder 38
Blohm & Voss Bv 138
  116
Blot 144
Blow Job 125
Blue Bandit 9
Bluebird 51
Blue Canoe 96
Bluehawk 30

INDEX

Blue Sentinel 96
Blue Thunder 98
BOAC 22, 63, 80, 94,
103, 119, 146
Boadicea class 103
Boeing 36, 46, 131, 173
B-17 Flying Fortress
54
B-29 Superfortress
145, 162
B-52 Stratofortress
91
C-135 Stratolifter
118
CH-46 Sea Knight 55
CH-47 Chinook 150
E-3 Sentry 73
E-6 Mercury 73
KC-135 Stratotanker
118
Model 247 85
Model 360 101
Model 377
Stratocruiser 85,
125
Model 707 2, 51, 72,
85
Model 720 85
Model 727 51, 133
Model 737 8
Model 747 3, 51, 74
Model 757 51
Model 767 51
P-26 98
XPBB-1 Sea Ranger
84
Bolingbroke 20
Boly 20
Bombcat 139
Bomber Destroyer 91
Bombphoon 135
Bomfire 117
Bone 114
Bosbok 19
Boulton Paul
P 3 Bobolink 64
P 64, P 71A 103
P 82 Defiant 39
P 108 Balliol 104
P 120 19
Box 51
Boxcar, Flying 11
Box Kite 79, 110
Brab 119
Brancker, Sir Sefton
105
Brasshat 110
Breguet
Bre 1 34
L1, L2, G3 147
Model 14 81
Model 763 104
Brewster
F2A Buffalo 13
SB2A Buccaneer 20
Brick 11
Brick, Flying 7
Bristol
F2 Fighter 23
M1 26
Scout 28
Type 95 Bagshot 21
Type 130 Bombay
15
Type 142M Blenheim
20
Type 152 Beaufort
15
Type 156 Beaufighter
145
Type 164 Brigand
24

Type 167 Brabazon
119
Type 170 17
Type 171 Sycamore
27, 77
Type 173 77
Type 175 Britannia
146
Type 192 Belvedere
84
British Aerospace
ATP 25
BAe 125 72
BAe 146 51, 74
BAe 748 24
Harrier, Sea Harrier
74, 93
Hawk 133
Jetstream 25, 126
SABA 93
Britten Norman
BN-2 Islander,
Trislander 50
BN-3 Nymph 94
Brittle 71
Brown Bomber 147
Brownie 64
Bryza 112
Bucc 11
Buchon 7
Bucker
Bu 131 Jungmann
154
Bu 181 Bestmann 57
Old Bucket Seats 59
Buffalo 10
Bullet Bomber 1
Bullfinch 26
Bullshit Bomber 58
Bumble Bee 148
Bump 7
Bus 79
Bushmaster 2000 136
Bushranger 68
Bustling Walter 13
Buzz Bomb 109

Cab 32
Cabin Cruiser 51
Cadillac 1
Caesar 2
Cali 136
Cambrian Airways 59
Camel 12, 66
Canadair
CL-41 Tutor 132
CL-44 146
CL-215 23
Canadairbus 146
Canadian Armed Forces
29, 41, 45, 55, 61,
68, 73, 93, 95, 108,
125, 131, 146
Canadian Car &
Foundry CBY-3
Loadmaster 18
Canadian Pacific 41
Can Opener 21
Canso 29
Canuck 72
Cap 38
Caproni; Ca 313 34
Caraja 71
Cardington Kite 86
Cargoliner 47, 62
Cargo Liner 17
Cargomaster 60
Cargonaut 139
Caribou 83
Carioca 31
Carioquina 31
Carrot, Flying 83

Carvair 94
CASA
C-101 Aviojet 62
C-212 Aviocar 100
Cassiopée 14
Caterpillar 37
Cathedral, Flying 124
Cat's Cradle 80
Caudron C 272 84
Cavalier 5
Centaur 70
Centaurus 81
Centennial 47
Ceres 100
Cervino 122
Cessna 26
CH-1 114
L-19, O-1 Bird Dog
92
L-27, U-3 21
Model 120, 140 140
Model 172 110
Model 208 Caravan 1
28
Model 210 Centurion
111, 123
Model 303 Crusader
33
Model 310 21, 111
Model 337 Skymaster
90
Model 340 111
Model 402 139
Model 404 Titan 28
Model 414 Chancellor
111, 139
Model 421 111
Model 425 Conquest 1
36
Model 500 Citation
49
O-2 90
T-37 139
UC-78 110
Challenger 71
Champagne Glass 128
Chance Vought (see
Vought)
Chantra 32
Chase C-123B 8
Cheesefighter 41
Cheetah 68, 80, 89
Chesapeake 149
Chetverikov Che-2 166
Chickasaw 71
Chicken Coop 116
Chickenhawk 12, 110
Chidori 113
Chimo 61
China
F-8 163
Qiang-5 79, 163
Chinese Scout 50
Chiricahua 64
Cierva 8, 111
W 11 Air Horse 123
Chocolate Lorry 64
Chummy Hearse 44
Chung Cheng 55
Cigar, Flying 95,98,149
City Jet 8
Clara 7
Clementine 44
Clog, Flying 116
Clothes Horse 123
Clunk 12
Clutching Hand 44
Clyde Clipper 18
Clydesdale 101
Coastguarder 25
Coastwatcher 35
Cobra 68

Cochise 9
Clockwork Mouse 151
Codfish 99
Coffee Pot 81
Coffin 124
Coffin, Flying 26, 44,
54
Colemill Enterprises 9,
21, 39, 71
Collegian 14
Colossal Guppy 91
Comanchero 71
Combat Scout 78
Combat Shadow,
Combat Talon 65
Comfy Levi 65
Comic 28, 52
Commando 113
Commonwealth
CA-4 Wackett
Bomber, CA-11
Woomera 144
CA-12, CA-13, CA-19
Boomerang 22
CA-15 75
Wirraway 22, 100
Commonwealth
class 40
Commuter 1, 106, 138,
147
Compass Call 65
Conair 68, 125, 139
Condor 143
C-One-Oh-Boom 11
Conestoga 20
Conger Eel 128
Consolidated (see also
Convair)
B-24 Liberator 11
B-32 133
Model 31 104
PBY Catalina 29,
166
Constancia II 75
Consul 96
Containership 74
Contra-Rotating Nissen
Hut 115
Convair (see also
Consolidated &
General Dynamics)
B-36 27, 98
B-58 Hustler 39
C-131 Samaritan 88
CV-240, CV-340, CV-
440, etc 51, 85, 88
CV-880 86
CV-990 51, 86
F-102 Delta Dagger
30, 39
F-106 Delta Dart 30,
119
L-13 70
R3Y Tradewind 84
XFY-1 102
Converter 20
Convoy 29
Corisco 31
Corn Cutter 115
Coronado 86
Coronet Solo II 65
Corrugated Coffin 8
Cosmopolitan 88
Cougar 135
Counter Invader 132
Coupe 24
Cow, Flying 65
Crab 6, 44
Crane 110
Crazy Water Buffalo
121
Crécy 149

Credible Hawk 30
Creek 78
Crossair 33, 74
Crouze 88
Crowd Killer 42
Crown 65
CUBy Sport, CUBy
Acro Trainer 53
Cunliffe-Owen OA-1
18
Currie Wot 145
Curtiss 173
C-46 Commando 35
F8C, O2C Helldiver
33
JN 72
NC 93
P-36 91
P-40 Warhawk,
Tomahawk,
Kittyhawk 91
SBC-4 Helldiver 33
SB2C Helldiver 15
SO3C Seagull 105,
113
XBTC-2 45
XF-87 19
XP-55 6
Cutlass 101

Daffy 39
Dagger 76
Daily Mirror Aircraft
80
Dak 60
Dakleton 60
Dakmaster 59
Dakota 58
Dantorp 77
Dart I 110
Dart II 140
Dash 3 125
Dassault
Alpha Jet 81
MD 450 Ouragan
137
MD 452 Mystère
(fighter) 43
MD 550 Mirage I 39
Mirage III 10, 76, 88
Mirage 5/V 76, 88
Mirage 2000 141
Mystère (corporate)
48
Dauntless II 121
David 2
Defender 69, 83
De Havilland (see also
Hawker Siddeley)
DH 2 122
DH 4 51
DH 6 44
DH 9 93
DH 10 4
DH 60 Moth 104,
105
DH 61 Giant Moth
28
DH 82 Tiger Moth
105, 135
DH 86 40
DH 89 Dragon
Rapide, Dominie
10
DH 95 Flamingo 65
DH 98 Mosquito 92
DH 100 Vampire
122
DH 104 Dove 40
DH 106 Comet
119
DH 108 130

INDEX

DH 110 Sea Vixen 110
DH 112 Sea Venom 5
DH 114 Heron 42
DH 125 72
De Havilland Canada 173
DHC-1 Chipmunk 31
DHC-3 Otter 125
DHC-4 Caribou 93
DHC-5 Buffalo 28
DHC-6 Twin Otter 140
DHC-7 Dash-7 107
DHC-8 Dash-8 58
D.F.W.
B I, B II 10
R I 56
Delfine 13
Delta Dud 40
Desert Hawk 30
Destroyer 62
Deuce 38, 43
Devil's Sled 52
Diamant 82
Digger's Delight 22
Dinky Toy 147
Diplomat 40
Dirt Eater 11
Discoverer 46
Discovery class 142
Disposalsydes 92
Dizzy Three 59
Dominie 10, 72
Doomsday Plane 119
Dora 7
Dora 9 21
Dornier
Alpha Jet 81
Do 11, Do 13, Do 23 34
Do 17 99
Do 24K 151
Do 28D-2 Skyservant 49
Do 335 5
Double 151
Double Breasted Club 110
Douglas (see also McDonnell Douglas) 173
AD Skyraider 121
A3D Skywarrior 145
A4D Skyhawk 12
A-26, B-26 Invader 132
B-18 Bolo 41
C-74 Globemaster, C-124 Globemaster II 25
DB-7 series 22,161
DC-2 39, 41, 159
DC-3 51, 58, 85, 159, 162
DC-4 51, 85, 94
DC-4E 85, 157
DC-6 33, 85
DC-7 51, 95, 115
DC-8 41
DC-9 93
F3D Skyknight 149
F4D Skyray 53
PD-808 142
SBD Dauntless 12
Dove 105
Dowager Duchess 59
Dragmaster 24
Dragon Express 40
Dragonfly 139

Dragonship 59
Dream of Elwin, The 59
Droopsnoot 53
Drut 149
Duck 59, 116
Duckbutt 57
Duck Nose 38
Dumodliner 47
Dumptruck, Flying 64, 121

Eagle 11, 49, 61, 123
Eaglet 21
Earthpig 1
Eastern Air Lines 36, 41, 43, 133, 135
Easy Rider 11
Edgley EA-7 Optica 25
Edsel, Flying 1
Egg 83
Eggbeater 32
EH Industries EH101 61
Eighty-Eight Day Plane 142
Eiko 26
Elation 36
Electric Fox 1
Electric Jet 143
Electric Wedge 143
Electric Whale 145
Electron 96
Elephant 43
Elephant Joyeux 71
Elizabethan 83
El Tomcat 57
Ely 700 14
Embraer 31, 70, 71
EMB-110 Bandeirante 12
EMB-312 Tucano 118
Emil 7
Empire Class 146
ENAER, Queen Air 63, 89, 106
T-35 Pillan 130
English Electric
Canberra 37
Lightning 55
Enforcer 5
Ensign Eater/Eliminator 16
Eric 113
Erik 2
Eshet 77
Esquif 147
Esquilo 7
E-Type Helicopter 110
Eule 47
Eurocopter (see also Aerospatiale & MBB)
AS 350 Ecuriel 6
AS 365 Dauphin 42
SA 330 Puma 135
Eurofighter 2000 69
Europa Jet 133
Eversharp 99
Excalibur 36, 55, 114, 131
Exec-Liner 141
Executive 122
Executive 400 40
Executive 600 21
Expediter 106
Expeditor 46
Express Air Liner 41

Fairchild

A-10 Thunderbolt II 93, 144
AU-23A Peacemaker 64
C-82 Packet 41
C-119 Flying Boxcar 41
C-123 Provider 8
FH-227 139
PT-19 Cornell 37
T-46 134
UC-61 6
Fairey (UK & Belgium) 52
IIIF 105
Albacore 5
Barracuda 13
Battle 47
Fantôme 50
Fox 75
Long Range Monoplane 103
Swordfish 127
Faithful, Old 121
Fan Jet 400 141
Fan Jet 600 147
Fantail 123
Farman
1909 Biplane 79, 105
F 40 66
MF 7, MF 11 116
Fastback 24
Fat Bertha 133
Fat Face 121
Father & Son 90
FBA (Franco-British Aviation) Type B 10
Fennec 7
Ferret 119, 148
Fertile Myrtle 13
FFA (Switzerland)
AS202 Bravo 150
C-3605 Schlepp 101
P-1604 72
Fiat (see also Aeritalia)
BR 20 Cicogna 158
C 6B 82
CR 32 32
G 50 50
Fieseler
Fi 103 (V-1) 109
Fi 156 Storch 38
Fifi 13
Filip 2
Films & TV
Battle of Britain, The 7
Blue Thunder 98
Cone of Silence 129
Flight of the Phoenix, The 42
Iron Maiden, The 63
No Highway 63
Sound Barrier, The 104
Finger 76
Firecat 125
Firefighter 139
Firekiller 129
Firestar 96
Fishington 149
Flagship 88
Flakbeau 146
Flaming Coffin 109
Flapjack 97
Flash 67
Flatiron, Flying 103
Flat Iron 2, 4
Flicker 67
Flicknife 21

Flight magazine 3, 20, 81
Fluff 9
Flutterschmitt 7
Fly 108
Flying Tiger Line 41
FMA (Argentina)
IA 35 Huanquero 75
IA 58 Pucara 39
Focke-Achgelis Fa 223 Drache 67
Focke-Wulf 173
Fw 58 Weihe 124
Fw 187 48
Fw 189 47
Fw 190 27, 90, 155
Fw 191 104
Fw 200 Condor 80, 159
Heuschrecke 111
Ta 152 90
FOD Vacuum 60
Foil 67
Fokker (German & Dutch)
50 138
100 51
D VIII 108
F VIIB/3m 133
F 27 Friendship 138
G I 62
Folland
Fo 141 Gnat 111
Type 43/37 55
Fongshu 82
Ford 12
Ford Tri-Motors 136
Ford's Folly 11
Forty-Four 146
Forty-Niner 86
Foster-Wickner Wicko GM1 144
Fouga Magister 146
Fouga 90 147
Foxstar 9
FRED 53
Freeman's Polly 92
Freightliner 147
Freightship 138
Frisbee 73
Fritz 7
Frog 68
Frying Pan 128
F&W (see FFA)

Gabriel 102
Gama 97
Gannet 1, 60
Gas Ring, Flying 115
'Gator 9
Gaviao 80
Gazelle 94
G-Car 31
Gekko 41
Gemini 140
Gemini ST 79
Gemisbok 135
General Aircraft Ltd (UK) GAL55 138
General Dynamics
A-12 Avenger II 99
F-16 Fighting Falcon 132, 143, 148
F-111 1, 148
Genet 144
Gentleman's Swordfish 5
Geofiz 82
Georgia Bel 24
Geronimo 140
Ghost 150
Gipsy Rose Lee 91

Giraffe 46
GLOB 118
Globe GC 1 Swift 24
Gloster
E 1-44 60
E 28/39 101
Gamecock 80
Javelin 1
Meteor 87
TC 33 60
Goblin 13
Go-Go Bird 150
Golden Arrow 86
Golden Falcon 41
Golden Football 88
Golden Jet 73
Gomershark 55
Gondola 111
Goofington 149
Goshawk 133
Gosport 6
Gotha G I 138
Governeur 80
Government Aircraft Factories (GAF)
Jindivik 100
Nomad 58
Graduate 30
Granville Bros 6
Grasshopper 115
Green Lizard 91
Grey, C.G. 26, 99, 116
Grey Ghost 19
Grey Old Lady 115
Grid 79
Griffon 68
Grob
G 103 Twin Acro 102
G 109B 102
G 115 65
Groundhog 66
Ground Mog 87
Groupe Technique de Cannes 14, 16
Old Growler 100
Growler 115
Grumman 10, 173
A-6 Intruder 72
AF Guardian 50
E-2 Hawkeye 97
EA-6B Prowler 72, 149
EF-111A Raven 1, 149
FF-1, F2F, F3F 13
F4F Wildcat 98
F6F Hellcat 1
F7F Tigercat 132
F8F Bearcat 16
F-14 Tomcat 132, 139
G-21 Goose 61
G-44 Widgeon 60
G-63 Kitten 64
G-159 Gulfstream 1 1
G-164 Ag-Cat 77
OV-1 Mohawk 146
SA-16 Albatross 57
S2F Tracer 125
TBF Avenger 32
WF-2 Tracer 125
X-29 102
Guardian 48
Guardrail 141
Guion 100
Guppy 33, 121
Guppy-type freighter conversions 91, 107, 125, 146
Gustav 7

INDEX

Gutless 45
Gyrone 101
Habu 18
Haddock 99
Haitun 42
Halton 63
Ham Bone 128
Hamilton Aviation 46, 50
Handley Page
HP 35 Clive 32
HP 42 95
HP 50 Heyford 44
HP 52 Hampden, HP 53 Hereford 128
HP 54 Harrow 122
HP 57 Halifax 63
HP 80 Victor 63, 141
HP 137 Jetstream 126
HPR 1 Marathon 33
HPR 3 Herald 97
O/100, O/400, V/1500 21
Hannover CL II, CL III 63
Hansa-Brandenburg D I 124
Harbin Yun-12 97
Harmonious Dragmaster 2
Harukaze 79
Harvard 100
Hauler 108
Havoc 22
Hawker 800, 1000 72
Hawker
Audax 64
Fury 67
Horsley 77
Hunter 2
Hurricane 69
Nimrod 67
P 1121 69
P 1127, P 1154 74
Typhoon 135
Woodcock 38
Hawker Siddeley
HS 125 72
HS 146 74
HS 748 24
Harrier/Sea Harrier 74, 93
Hawk 133
Trident 62
Head Shed 65
Heavyweight Glider 133
Hebrides class 42
Heinemann's Hot Rod 12
Heinkel
He 46 64
He 51 31
He 66 158
He 70 20
He 111 98, 154
He 112 7, 156
He 162 90, 144
He 177 Greif 109
Hei Ying 30
Heli-Camper 71
Heliliner 61
Helio H-250A Courier II 27
Helitanker 68
Helldiver 33
Helmore 22
Helo 32
Henderson HSF 1 61
Henschel

Hs 123 5
Hs 129 149
Hera 97
Hermes 73
Heuschrecke 111
Hibari 57
Hiller
FH-1100C 67
Model 360 108
Himalaya class 103
Hina-Zuru 96
Hispano
HA-200 Saeta 100
HA 1109, HA 1112 Buchon 7
HS-132-L 32
Hiss Two 113
Hiyodori 68
Hog 67
Hog Nose 16
Honeymoon 52
Horned Butterfly 80
Hornet 57
Horseshoe 16
Hose Nose 16
Hot Rod 39
Hudson 22
Hughes (see also McDonnell Douglas)
HK 1 Hercules 123
Model 269, Model 300 95
OH-6 Cayuse 83
XH-17 Flying Crane 54
Humiliator 55
Humpback 86
Humu 14
Hunchback, Devil's Chariot, Shturmovik 8, 40, 117
Huron 141
Hurry/Hurryback 70
Husky 36, 73
Huzar 5
Hyper Hog 133
Hydraulic Palm Tree 32
Hythe 103
Ibex 85
ICA (Romania)
IAR-99 Soim 130
IAR-317 Airfox 31
Ilyushin
Il-2 117,161
Il-4 161
Il-10 117, 161
Il-12 162
Il-14 112, 163
Il-18 91, 162
Il-20 69
Il-28 161, 162, 165
Il-38 165
Il-40 161
Il-54 161
Il-62 162
Il-76 55, 162
Il-78 165
Il-86 2, 162
Imperial Airways 40, 45, 95, 103, 112
Impossible 71
Incinerator, Flying 116
Incredible 71
Indian 25
Indian Air Force 8, 9, 31, 38, 55, 107, 116, 129, 137, 141

Infantryman, Flying 117
Infinite 47
Innovator 29
Interceptor 48
Intercity 748 25
Interstate S-1B Cadet 61
Intruder 37
Invader 5
Iron Annie 8
Iron Butterfly 133
Iron Gustav 117
Iron Maiden 63
Iroquois Warrior 120
Isfahan 68
Islander class 10
Israel Aircraft Industries
Arava 28
Westwind 34
Israeli Air Force 12, 43, 76, 140
Jackaroo 135
Jaeger 33
Jaguar 61
Japanese Self Defence Forces (post-war) 26, 55, 57, 68, 79, 96, 106, 131
Jaybird 100
Jayhawk 16, 30
Jeep, Flying 61, 64
Jet Cruiser 122
Jet Liner 40
Jet Profile 140
Jetstream 36
Jet Trader 41
Johan 2
Johnson, Lyndon B. 3, 18
Jolanthe the Pig 127
Jolly Green Giant 113
Jumbo Jet 41
Jungly 66, 113
Junkers
G31 125
J1 136
J1 55
Ju 52/3m 8, 159
Ju 87 127, 156
Ju 88 90, 133, 156
Ju 188 133
Ju 268 90
Ju 287 90
Ju 290 95
Ju 322 Mammut 57
Ju 388 133
Jupiter class 8
Kahu 12
Kalkadoon 10
Kaman
H-2 Seasprite 66
H-43 Huskie 98
Kamikaze Aircraft, Wakazakura, Baka 56, 144, 154
Kamov
Ka-10M 164
Ka-15 164
Ka-18 165
Ka-20 164
Ka-22 Vintokryl 165
Ka-25 164, 165
Ka-26, Ka-126 165
Ka-27, Ka-29 164
Ka-50 Werewolf 165
Kangaroo 27, 86
Kansan 46
Kattercopter 78

Kauz 99
Kawanishi
E7K 154
E15K Shiun 157
H3K 154
H6K 157
H8K 155
N1K Kyofu 158
N1K Shiden 155
Kawasaki
BK 117 121
Ki-10 158
Ki-32 157
Ki-45 Toryu 157
Ki-48 156
Ki-56 159
Ki-61 Hien 159
Ki-64 158
Ki-102 158
Type LO 159
K-Car 31
Kestrel 74
Kiddie Kar 122
Killer 148
King Beaver 125
Kingbird 138
KingCobra 120
Kitty 91
Kittyhawk 91
Kitty Hawk 76
Klemm L 25 130
Klong Bird 65
Kokusai
Ki-59 159
Ki-76 158
Ki-86 154
Ku-7 154
Ku-8 155
Koliber 44
Kolkhoznik 82
Kudu 20
Kurnass 43
Kusho H7Y 159
Kyokko 41
Kyushu
K10W 157
Q1W Tokai 157
Labrador 55
LaGG (see also Lavochkin) LaGG-3 61
Lala-1 82
Lamp Lighter 11
Lancastrian 80
Lancer 26
Landbee 134
Landseaire 29
Lanky 80
Laser 114
Last of the Gunfighters 88
Lavochkin (see also LaGG)
La-7 163
La-9 164
La-11 163
La-15 163
La-160 6
La-250 4
Lawn Dart 143
Leaping Heap 75
Learjet 5, 137
Learstang 5
Leisure Jet 74
Lemon 1
Leone 54
Lesnik 115
Leukoplastbomber 124
LFG (see Roland)
Lib 11
Liberty Plane 51

Lifter 84
Lifting Body aircraft 124
Liftmaster 33
Lift Ship 46
Lighter, Flying 95
Lightning Bug 68
Limping Annie 48
Lion 77
Lisunov Li-2 162
Little Ack 17
Little Critter 111
Little Donkey 108
Little Hawk 77, 108
Little Liner 46
Little Red Riding Hood 118
Little Tiger 54
Little Woodpecker 127
Lizard 1
Loch class 22
Lockheed 141, 174
Aurora 105
C-5 Galaxy 53
C-130 Hercules 64
C-141 StarLifter 24, 128
Constellation 36
Electra (Model 10) 22
F-16 Fighting Falcon 132, 143, 148
F-22 81
F-104 Starfighter 26, 30
F-117 150
Hudson 22
JetStar 3, 49
LASA-60 20
Lodestar 81, 159
Orion 108
P-2 Neptune 96
P-3 Orion 96
P-38 Lightning 53
S-3 Viking 66
SR-71 18
Super Electra 22, 159
T-33 131
TriStar 135
U-2 43
Vega 102
XP-58 Chain Lightning 151
Locust 21
Loening OL, OA-2 123
London Bus 63
Long Bird 95
Longhorn 116, 137
Long-Nosed Dora 27
LongRanger 76
Longreach 74
Loon 109
Lo Presti Piper
SwiftFire, SwiftFury, SwiftThunder 24
Luftwaffe's Lighter 109
LuxuryJet, LuxuryLiner 51
L.W.F. Model V 84
Lynx 90
Macchi MC205 Veltro 95
Mach 1 147
Machete 39
Magic Roundabout 115
Magnesium Monster/Cloud 98
Makalu 147

INDEX

Mammoth Water Spider 54
Mansyu Ki-71 155
Mantis 94
Mara 89
Marine One 3
Maritime Enforcer 138
Marketeer 132
Marksman 132
Marquis 9
Martian 148
Martin
  AM Mauler 1
  B-26 Marauder 148
  B-57 37
  M-130 31
  Mars 23
  Model 167W 86
  Model 2-0-2 46
  X-24 103
Martinsyde
  F3, F4 Buzzard 92
  G 100, G 102 45
Martlet 98
Master 14
Matador 75
Maule M-4 38
Mayfair class 136
MBB (Messerschmitt Bolkow Blohm)
  BK 117 121
  Bo 105 45
McCudden, James 20, 104, 116
McDonnell & McDonnell Douglas
  A-12 Avenger II 99
  AV-8 Harrier 74
  DC-10 47, 51
  F/A-18 Hornet 34, 132
  F2H Banshee 12
  F-4 Phantom 43, 148
  F-15 Eagle 132, 140, 148
  F-101 Voodoo 30, 95
  MD-11 51
  MD-80 51
  Model 500 83
  T-45 Goshawk 133
  XF-85 Goblin 27
  XP-67 91
  YC-15 73
McNamara's Folly 1
Mechanical Cow 116
Mega Top 74
Merchantman 62
Mercurius 72
Mercury 73
Merlin 30, 106
Merpati 72
Merseburg 57
Mescalero 110
Messerschmitt (see also MBB)
  Bf 108 Taifun 4
  Bf 109 7, 90, 157
  Bf 110 152, 155
  Me 163B Komet 52
  Me 261 2
  Me 262 90, 112
  Me 264 4
  Me 410 67
  Me 321, Me 323 Gigant 124
Meteor 144
Methuselah, Old 59
Metro 106
Mewa 140
M.F.I. (Sweden) MFI-9 Junior 89

MiG (see Mikoyan & Gurevitch)
MiG 5 55
Mighty Ear Banger 128
Mighty Munk 32
Mignet HM 14 Pou de Ciel 52
MiG (Mikoyan & Gurevitch)
  MiG-9 163
  MiG-15 60, 163, 165
  MiG-17 119, 164
  MiG-19 79, 163
  MiG-21 9, 163, 166
  MiG-23, MiG-27 38, 164
  MiG-23DPD 163
  MiG-25 110, 164
  MiG-29 8, 164
  MiG-31 164
  'MiG-37' 163
  Ye-152A 164
Mil
  Mi-1 92, 164
  Mi-2 75, 165
  Mi-4 130, 165
  Mi-6 165
  Mi-8 107, 112, 165
  Mi-10 164
  Mi-12 37, 165
  Mi-14 164
  Mi-17 107, 165
  Mi-24 40, 164
  Mi-26 164
  Mi-28 164
  Mi-34 164
Miles
  LR5 22
  M3E Falcon Six 56
  M9A Master 76
  M14 Magister 84
  M25 Martinet 106
  M38 Messenger 85
  M52 56
  M60 Marathon 33
  M68 Boxcar 75
  M100 Student 30
Mili-Role 90
Milkbottle, Flying 73
Mini-COIN 89
Mini-Tanker 46
Minuano 31
Miscellaneous, Old 121
Missile with a man in it 26
Mistral 122
Mitchell, Billy 13, 18
Mithras 43
Mitsubishi 96
  A5M 154, 158
  A6M 151, 154, 158, 159
  A7M Reppu 158
  B5M 157
  C5M 154
  F1M 158
  G3M 157, 159
  G4M 95, 144, 154
  J2M Raiden 156
  J4M 157
  J8M1 Shusui 156
  K3M 158
  Ki-2-II 157
  Ki-21 158
  Ki-30 154
  Ki-46 155
  Ki-51 158
  Ki-67 Hiryu 158
  Ki-73 158
  L3Y 159
  L4M 159
  Mu-2 47

Mu-300 Diamond 15
TK-4 155
Mizar 90
Monarch 26 132
Mongoose 12, 98, 148
Moonfighter 22
Morane, Morane Saulnier
  MS 406, MS 410 91
  MS 500 38
  MS 880 Rallye 44
  Type N 8, 26
Mortician's Mate 61
Mosquito 100
Mother 61
Mother Sow 65
Mouse Catcher 116
Movies (see Films)
Mower 62
M'Pala 94
Mraz K-65 Cap 38
Mudguard 62
Mud Hen 140
Mule 7
Murderer 148
Mushshak 89
Myasischev
  Mya-4 Molot 7, 161
  Mya-50 161
  M-17 166
Mystery Jet 122

Nakajima
  A6M2-N 158
  AT-27 155
  B5N 156
  B6N Tenzan 156
  C6N Saiun 157
  E2N 154
  E8N 155
  G5N Shinzan 157
  G8N Renzan 158
  J1N Gekko 156
  Ki-27 157
  Ki-43 Hayabusa 157
  Ki-44 Shoki 159
  Ki-49 Donryu 155
  Ki-84 Hayate 155
  L1N 159
Nammer 77
'Nana 11
Narvik Nightmare 117
Natacha 108
Navigator 46
Nellie 150
Nesher 76
New Zeland Aerospace CT-4 Airtrainer 102
Nieuhawk 85
Nieuport
  Model 11 15
  Model 17 118
Nieuport & General Nighthawk 85
Nightfox 84
Nighthawk 23, 37, 68
Night Hawk 30
Night Intruder 37
Night Owl 78
Night Ranger 78
Night Stalker 48
Nikko 79
Nine Hour Cruiser 51
Nineteen 48
Nisr 44
Nite-Writer 71
Noisy Star 94
Nomad 29, 50, 108
Nomair 50
Nord
  Griffon 62

Model 262 91
Noralpha, Pingouin 4
Norman Aircraft Company NAC 1
  Freelance 94
Norn 67
North American (see also Rockwell)
  A-5 Vigilante 143
  AJ Savage 112
  B-25 Mitchell 18, 161
  F-86 Sabre 41, 111
  F-100 Super Sabre 30, 68, 148
  F-107 30, 129
  F-108 Rapier 30
  FJ Fury 41
  P-51 Mustang 5
  T-6, SNJ Texan 100, 157
  T-28 Trojan 50
  T-39 Sabreliner 41, 148
Northrop 174
  A-17A 94
  B-2 Spirit 14
  F-5 54
  F-18L 34
  F-20 Tigershark 55
  F-89 Scorpion 60
  M2-F2 27
  Model 3A 151
  P-61 Black Widow 19, 110
  P-530 34
  T-38 Talon 147
  TR-3A 19
  XP-56 19
  XP-79B 107
  YC-125 Raider 101
  YF-17 34, 132
  YF-23A 19
Norwegian class 103
Nuri 113

Observer 53
Ocean Hawk 30
Opus 280 66
Oryx 135

Pablo 98
Packplane 42
Pan Am 9, 31, 33, 48, 74, 115
Panavia Tornado 21
Panda 71
Pandora 23, 75
Panhandle, Flying 128
Pantera 89
Panther 21, 42, 71
Partenavia Oscar 48
Pathfinder 53
Pathfinder 206 78
Patrulha 12
Pave Hawk 30
Payloader 113
Pave Low III 30, 127
P-Boat 29
Peacemaker 64
Peanut Special 13
Peeled Banana 11
Peenemunde 20 109
Peenemunde 30 52
Peeping Tom 139
Peewit 104
Pegasus 101, 140
Pelican 57, 113
Pemberton Billing (see Supermarine)
Penetrator 68

Perce 104
Percival
  Mew Gull 16
  Prince 98
  Proctor 104
  Q6 99
Peter-Dash-Flash 5
Petlyakov Pe-2 99, 162
Petrel 34, 61, 106
Petulant Porpoise 60
Phalcon 73
Phantom Diver 87
Phoenix 71, 108, 129
Piaggio
  P 7 128
  P 136 111
  P 149 101
  P 166 111
  PD-808 142
Piasecki
  H-21 Shawnee, Workhorse 10
  HRP-1 Rescuer 10
  HUP Retriever 69
Piccolo Tube 119
Piel Emeraude 82
Pig, Flying 142
Pig 1, 8, 40, 137
Pig Boat 29
Pilatus PC-6 Turbo Porter 64
Pinguoin 4
Pionair 59
Piota 59
Piper (see also Lo Presti Piper) 25, 174
  Aerostar 114
  J-2, J-3 Cub 52, 61
  PA-15 Vagabond 144
  PA-18 Super Cub 53
  PA-23 Apache, Aztec 140
  PA-28 31, 130
  PA-31 Navajo, Chieftain, Mojave 70
  PA-32 Cherokee Six 31
  PA-34 Seneca 140
  PA-48 Enforcer 50
Placid Plodder 59
Plain Jane 118
Platypus 35, 37, 49
Plebe 24
Plymouth class 103
Pocket Fighter 111
Polaris 45
Polikarpov
  I-15 Chaika 120
  I-16 108
  Po-2 15, 115, 166
  R-5, R-Z 108
Po One 36
Poor Man's Grummans 13
Pop-Up Target 78
Porcupine 119
Porker 145
Portofino 111
Possum 57
Pot Hawk 30
Power Egg 52
P-Plane 109
Pratap 107
Praying Mantis 45
Preceptor 104
Prefect 37
Pregnant Beast 32
Pregnant Dakota 136
Pregnant Guppy 126
Pregnant Pencil 99
Pregnant Perch 83

INDEX

Pregnant Pig 81
President 99
President 600 9
Prestwick Pioneer 100
Privateer 11
Prospector 53
Prostitute, Flying 148
Protector 72
Protruder 72
Pterodactyl 45
Pudgie 74
Puffship 59
Puff, the Magic Dragon 59
Pulpit 107
Pulveriser 97
Push-Me-Pull-You 90
Push-Proj 121
PZL (Poland) 76, 112, 140
  PZL-104 Wilga 56
  PZL-110 Koliber 44
  TS-11 Iskra 92
  W-3 Sokol 4, 112

Qantas 40, 73, 74
Qiang-5 79, 163
Queer Bird 105
Quick Fix I 68
Quick Fix II 30
Quickie Aircraft Free Enterprise 17
Quickstrike 139
Quiet One, The 84
Quiet Trader 74

R'am 76
Racher 133
Rackdan 77
Radome 36
Rakshak 38
Ramier 4
Rampage 87
Ramp Rooster 63
Ranger 22, 57
Rapide 10
Raven 1
Reaper 87
Red Bandit 119
Red Flag 11, 55
Regal Beagle 14
Regent 1500 111
Reggiane Re 2000 Falco 1 64
Reid & Sigrist
  RS 1 100
  RS 3 Desford 21
Reims Aviation 28, 110
Reindeer 63
Republic 174
  F-84 Thunderjet 66
  F-84F Thunderstreak, RF-84F 97
  Thunderflash 128
  F-103 30, 134
  F-105
  Thunderchief 30, 133, 148
  JB-2, Loon 109
  P-47 Thunderbolt 73
  RC-3 Seabee 134
  XF-84H 128
Rescuemaster 94
Retaliator 143
Rhino 43
Rich Peter 13
Rivet Rider 65
Rock, The 150
Rockbeau 146
Rockwell
  B-1 Lancer 114
  Jet Commander 35

Model 112, 114 84
Space Shuttle Orbiter 23
T-2 Buckeye 7
Thrush Commander 143
Rodent 62
Roland (LFG)
  C II 145
  D II 116
Roleur 99
Rotorcoach 84
Royal Aircraft Factory
  BE2 107
  BE4, BE8 20
  BE9 107
  BE10 56
  FE2 49
  RE8 63
Royal Flying Corps 20, 26, 44, 45, 49, 63, 107, 116, 118, 122, 141, 147
Royal New Zealand Air Force 12, 21
Rumpety 116
Rumpler Taube 71
Rutan
  Model 151 ARES 93
  VariEze 33
  Voyager 18
Ryan
  FR-1 Fireball, XF2R-1 39

Saab
  J-29 39
  J-35 Draken 2
  JA-37 Viggen 45
  MFI-15 Safari 89
  Model 105 52
  Model 210 2
  SF340 33
Sabre II 10
Sabredog 41
Safari 89
Sage Type 2 23
Sagittario 54
Sahara 105
Saint-Exupéry, Antoine 81
Salamanda 4
Salamander 144
Samaritan 88
Sandringham 103
Sandy 120, 121
Sardine, Flying 31
Saro (see Saunders Roe)
Saunders Aircraft Corp
  ST-27, ST-28 42
Saunders Roe 87
  Lerwick 100
  Scout 114
  SR-N1 36
Savoia-Marchetti
  SM-75, SM-82 86
  SM-79 69
Sawn-Off Spitfire 104
Scarier 75
Scheibe SF250B Motor Falke 102
Schleicher ASK-21, ASW-19 102
Scottish Aviation
  Bulldog 26
  Jetstream 126
  Pioneer, Twin Pioneer 100
Scourge of the Atlantic 80
Scout 114

Scoutmaster 25
Scraper 108
Scrapper 50
Screaming Green & Brown Trash Hauler 64
Sea Dragon 127
Sea Eagle 30, 33
Seafang 117
Seafire 117
Seaford 103
Seagull 113, 116
Sea Mare 29
Seapig 1
SeaRanger 78
Sea Scan 35
Seattle 54
Seawolf 68
Seeker 23
Seely 24
Seminole 106, 131
Semiquaver 92
Senior Hunter 65
Sentinel 125, 138
Sentry 73, 90
Sepecat Jaguar 116
Sertanejo 31
Seversky 2PA 155
Shackle Bomber 115
Shadow 42, 123
Shady Lady 43
Shahbaaz 89
Shaky, Old 25
Shar 75
Sharp Nose 77
Shed, Flying 51
Sherpa 23, 28
Shirasagi 55
Shithook 150
Shoebox 23, 51
Shoehorn, Flying 123
Shorts
  330, 360 23
  Belfast 24
  Crusader 38
  R24/31 79
  S17 Kent 112
  S23 45
  S47 Triple Tractor 50
  SC1 16
  Silver Streak 129
Skyvan 51
Sperrin 141
Stirling 97
Sunderland 102
Tandem-Twin 57
Tucano 118
Showa L2D 159
Shrike 15, 27
Shuka 52
Shusui 52
Shute, Nevil 63, 96
SIAI-Marchetti SF-260 144
Siebel Si 204 86
Siemens-Schuckert
  DDr I 45
  R I to R VII 56
Sigrist Bus 52
Sigurd 2
Sikorsky (USA & Russia)
  HH-53 Super Jolly 127
  Ilya Mouromets 60
  R-4 67
  S-21 Russkii Vitiaz 60
  S-43 9
  S-51 43, 77
  S-55 71, 77

S-56 38
S-58 66
S-61 3, 112
S-64 Skycrane 129
S-65 Sea Stallion 127
S-76 123
S-80 Super Stallion 127
UH-60 Blackhawk 3, 29
VS-300 70
VS-316 67
Silver 113
Silver Dollar 69
Silver Eagle 123
Silver Sixty 132
Silver Star 131
Sirrius 44
Sister Gabby 58
Six Thousand Pound Dog Whistle 139
Skoda 142
Skoshi Tiger 54
Skybarge 29
Skybolt 10
Sky Cruiser 74
Skyfox 131
Skyhog 12
Skyhook 44
Sky Knight 95
Skylark 86
Skyliner 42, 47, 51, 59
Skymonster 146
Skypallet 42
Sky Pearl 13
Sky Scooter 78
Skysleeper 58
Skystreamer 88
Skytrain II 93
Skytruck 42
Skyvan 42
Sky Zephyr 22
SLAT 145
Sled 18
Slick 67, 96
Slick Chick 69
Slumberland 103
Smaragd 82
Smokey, Old 43
Smoking Joe 86
SNCASE 6, 71, 122
SNCASO
  SO-30 Bretagne 16
  SO-90, SO-94, SO-95 14
Sneeb 46
Snoopy 1
SNUF 75
Sociable 139
Sofia 76
Soldier's Aircraft 61
Solenoid, Flying 97
Solent 93, 103
Sonofabitch, 2nd Class 15
Sopwith 174
  1½ Strutter 52
  Churchill 139
  Dolphin 20
  L.R.T Tr 45
  Snipe 43
  Triplane 138
  Two-Seat Scout 123
  Type 806 62
Sioux 57
Spacemaster 41
Spad 40
SPAD SA 1 to SA 4, SG 1 107
Spamcan 26
Spanish Civil War 5, 13, 31, 32, 76, 98,

99, 108, 120, 127
Spark Plug, Flying 115
Spark Vark 1
Sparrowhawk 144
Spatz 144
Speaker Bird 27
Specter 150
Spectre 43, 65
Spectrum-One 90
Speed Brake, Flying 65
Speedfreighter 36
Speedy D 12
Sperm 83
Spider 124
Spit, Spitter 117
Spiteful 117
Spooky 58
Sport 93
Spraymaster 32
Sprite 115
Squash Bomber 133
Squawk 12
Squeak 139
Squirrel 7
Squirt 69, 101
Squirter 149
Stability Jane 107
Stag 94
Stalingrad Type 109
'Stang 5
Stanley Steamer 60
Star 250 47
Star Jet 114
Starlizard 128
Star Wars Fighter 34
Statesman 74
Statuscruiser 125
Steam Chicken 122
Steam Pigeon 116
Steer 131
Steward Davis 29, 42
Stickleback 102, 149
Stinger 42
Stortebeker 133
Strainer 125
Strategic Boomerang 14
Stratobladder 118
Stratoboozer 125
Strikefighter 120
Strike Pig 9
Sturmvogel 112
Submarine 119
Sud Aviation (see also Aerospatiale)
  Caravelle 67, 85
  Concorde 35
  SE 313 Alouette II 80
Sukhoi
  Su-2 117
  Su-7 163, 166
  Su-9, Su-11 163, 165
  Su-15 164
  Su-17, Su-22 86, 163
  Su-24 163
  Su-25 111, 164
  Su-27 37, 164
T-3 9
Sun 102
Sundowner 32, 93
Sunjet 73
Sunriser 23
Super 6 33
Super 7 10, 115
Super 26 132
Super Airbus 74
Super Bird 127
Super Transporter 107
Super Bird 127
Super Dog 69
Superdragon 74
Super Five 46

185

INDEX

Super Flipper 107
Super Fox 12
Superfreighter 17
Super Handley 21
Super Lifter 129
Super Liner 46
Superman 73
Supermarine (inc.
Pemberton-Billing)
Attacker 72
PB 9 115
PB 23E 121
Sea Otter 124
Southampton 93
Spitfire 116
Stranraer 125
Swift 104
Type 224 117
Type 322 44
Walrus 115
Super Mustang 136
Super Q 73
Supershed 23
Super Shitter 127
SuperStar 7
SuperStar 81, 114
Super Star 36
Super Stuka 16
Super Transport 68
Super V 39
Supporter 89
Surveiller 9
SUWACS 56
Swearingen 106, 131
Sweetheart of Okinawa 16
Swift 86
SwiftFire, SwiftFury, SwiftThunder 24
Swinger 21
Switchblade 1

Tachikawa
Ki-9 158
Ki-17 154
Ki-36 156
Ki-54 156
Ki-70 154
Ki-74 157
Tadpole, Flying 128
Tadpole 94
Taj Mahal, Flying 3
Tank, Flying 117
Tank Cracker 127
Tank, Kurt 27, 47, 48
Tarhe 129
Tarpon 33
Tasman class 103
Tatzelwurm 90
Taurus 76, 141
Tawron 42
Taylorcraft (US & UK)
Auster 70
Model D 61
Plus D 70
T-Bird 151
T-Bird II 141
T-Cat 35
Technolog 76
Teeny Weeny Weasel 148
Temco 24, 136
Ten Minute Killer 53
TexasRanger 78
Thomas Morse S-4 137
Thumper 68
Thunderblot 144
Thunderbolt 87
Thunderbug 109
Thunderjug 73
Thunderprop 128
Thunderscreech 128

Thunderthud 133
Tigereye 54
Tigershark 55
Tinfoil Bomber 80
Tin Lizzie 136
Tinsydes 45
Tin Triangle 3
Tin Whistle 147
Tit, Flying 147
T-Jet 133
Tomahawk 66, 91
Tomb 43
Tommo 91
Top Gun 12, 139
Toothpick Maker 91
Torbeau 146
Tornado 135
Torpedo, Flying 148
Torpedo 82
Toucan 8, 25
Tourer 24
Tourist 101
Tradewind 47
Trainer aircraft 151
Tramcar, Flying 60
Trannie 102
Transall C 160 102
Transavia PL12 M300 85
Transporter 28
Traveller 124, 144
Triangle, Flying 2
Tri-Commutair, Tri-Commuter 51
Tripod 62
Triton 36, 58
Tri Turbo-3 59
Trojan 20
Trooper 57
Troopmaster 136
Trug 2
Tsetse 92
Tsibin Ts-25 165
Tsiklon 91
T-tailed Mountain Magnet 128
Tub 151
Tupi 31
Tupolev
SB-2 76
Tu-2 161
Tu-4 162
'Tu-10' 161
Tu-14 161
Tu-16 118, 161
Tu-22 161
Tu-22M 161
'Tu-26' 161
Tu-28 163
Tu-70 162
Tu-82 162
Tu-85 161
Tu-91 131, 161
Tu-95 161
Tu-98 161
Tu-104 118, 162
Tu-110 162
Tu-114 162
Tu-124 162
Tu-126 166
Tu-134 163
Tu-142 161
Tu-144 9, 35, 162
Tu-154 162
Tu-160 161
Turbine Tanker 25
Turbo 18 46
Turbo 34 90
Turbo 67R 59
Turbo Commuter 59
Turbo Express DC-3 59

Turbo-Liner 88
Turboliner 46
Turbosky 49
Turbo Three 59
Turkey 32
Tweety Bird 139
Twinbee 134
Twin Cat 77
Twin Harvard 46
Twin Jet 46
TwinStar 7
Twin Sticker 151
Two-and-a-Bit Plane 27
Twopenny Trident 51
Twosader 88
Tzukit 147

U-Bird 16
U-Boat 65
Ugly Hooker 51
Uhu 47
Ultra Hog 133
Umibato 106
Unfinished 57
United Air Lines 85
Universal Freighter 124
Universel 104
Uragan 99
US Coast Guard 30, 41, 42, 48, 113
Useless 78, 110

Vacuum Cleaner 57
Valiant 102, 117
Valmet L-70 Vinka 90
Vanguard 102, 142
Varnished Guaranteed Coffin 61
Ventellation 81
Venture 102
VFW-Fokker VFW-614 48
Viceroy 96
Vickers
ES2 26
FB5 62
FB12 122
FB16D 104
FB19 26
No. 26 105
Type 432 136
Valetta 137
Valiant 18, 141
Vanguard 62
VC10 142
Victoria & Valentia 142
Viking 136
Vildebeest 15
Vimy 142
Vincent 15
Virginia 56
Viscount 142
Vulcan 100
Wellington 149
Victor 117
Viewmaster 59
Vigilant 102
Vijay 38
Viking 102
Vikram 9
Viper 120
Virus 56
Vista Liner 140
Vistaplane 71
V-Jet 73
Voisin LA 143
Volant Solo II 65
Volpar 46, 126
Vought

A-7 Corsair II 119
F4U Corsair 16
F7U Cutlass 45
F8U Crusader 88
SB2U Vindicator 149
V-143 151
XC-142 43
XF5U 97
Voyager 46
Voyageur 55
Vulcan 57
Vultee (see also Convair)
A-35 Vengeance 56
BT-13 Valiant 142
V-11GB 157

Waco CG-4A Haig 62
Wag-Aero Inc. 53, 144
Wakataka 131
Wallaby 93
Walrus 94, 113
War Hatchet, Old 91
Warning Star 36
Warrior 78
Wasp 57
Watanabe
E9W 158
K10W 157
Waterman, Waldo 6
Wayfarer 17
Wendover 83
Wessex 66
Westland
Belvedere 84
F20/27 Interceptor 36
Lynx 109
Lysander 83
Scout 114
Sea King 113
W30 49
Wallace 150
Walrus 94
Wapiti 150
Wasp 115
Wessex (helicopter) 66
Whirlwind (fighter) 37
Whirlwind (helicopter) 71, 77
Widgeon (helicopter) 43
WS-51 43
Westwind 35, 47
Whale 35, 119
Wheelbarrow, Flying 122
Whirlibomber 38
Whirligig 38
Whirlwind 71
Whispering Death 1
Whisperjet 133
Whisperliner 107, 135
Whistling Bird Cage 125
Whistling Death 16
Whistling Outhouse 98
Whistling Shitcan 75
Whistling Tit 147
White Bandit 80
Whitehouse, Flying 3
White & Thompson Type 1172 22
Wichita Wobbler 46
Wicked One 26
Widgeon 43
Widow Maker 26, 45
WIGE 28
Wildcatfish 98

Willy Fudd 125
Wiltshire 113
Wind Indicator 149
Window Breaker 100
Wirraway 100
Wisley 20
Wokka 32
Wolpertinger, Flying 49
Wombat 13
Wooden Wonder 61, 92

Xavante 70

Yakovlev
UT-2 165
Yak-3 77
Yak-7 165
Yak-8 163
Yak-9 77, 164
Yak-10 163
Yak-11 166
Yak-12 56, 163
Yak-14 165
Yak-15, Yak-17 163, 165
Yak-16 163
Yak-18 15, 165, 166
Yak-23 164
Yak-24 23, 165
Yak-25 164, 165
Yak-27 165
Yak-28 161, 162, 163, 165
Yak-30 165
Yak-32 165
Yak-36 164
Yak-38 164
Yak-40 150, 162
Yak-141 164
Yale 100
Yassour 127
Y-Boat 29
Yeoman class 17
Yokosuka
B4Y 156
D4Y Suisei 156
E14Y 155
H5Y 154
K5Y 159
MXY-7 Ohka 144, 154
P1Y Ginga, Byakko, Kyokko 155
Yukon 146

Zafar 300 78
Zeppelin Staaken R VI 56
Zilch 33
Zimmer Skimmer 97
Zinc 79
Zipper 26
Zoo Plane 3

186